R00081 48234

Ref
QE
882
C15
N94
1979

Cop.1

FORM 125 M

BUSINESS/SCIENCE/TECHNOLOGY DIVISION

The Chicago Public Library

Received   JUN 6 1981

NORTH AMERICAN QUATERNARY CANIS

Frontispiece

Three species of *Canis*. Top, *Canis lupus* (the gray wolf, photo by L. David Mech). Middle, *Canis rufus* (the red wolf, photo by Curtis Carley). Bottom, *Canis latrans* (the coyote, photo by Tom Smylie). All photographs courtesy of the Fish and Wildlife Service, Department of the Interior.

NORTH AMERICAN QUATERNARY *CANIS*

RONALD M. NOWAK

Staff Specialist
Office of Endangered Species
U.S. Fish and Wildlife Service
Department of the Interior
Washington, D.C. 20240

MONOGRAPH

OF THE

MUSEUM OF NATURAL HISTORY, UNIVERSITY OF KANSAS

NUMBER 6

1979

NORTH AMERICAN QUATERNARY *CANIS*

MONOGRAPH OF THE MUSEUM OF NATURAL HISTORY

THE UNIVERSITY OF KANSAS

Number 6, pages 1-154, text figures 1-55

September 1, 1979

Editor: E. O. Wiley

COPYRIGHTED

BY

MUSEUM OF NATURAL HISTORY

THE UNIVERSITY OF KANSAS

LAWRENCE, KANSAS 66045

U.S.A.

ISBN: 0-89338-007-5

PRINTED

BY

THE UNIVERSITY OF KANSAS PRINTING SERVICE

LAWRENCE, KANSAS

U.S.A.

CONTENTS

- INTRODUCTION ... 1
  - ACKNOWLEDGEMENTS ... 2
  - METHODS ... 4
  - AGE AND SECONDARY SEXUAL VARIATION ... 6
- HISTORY AND STATISTICAL ANALYSIS OF RECENT POPULATIONS ... 7
  - COMPARISON OF KNOWN SERIES OF DOGS, WOLVES, AND COYOTES ... 7
  - SYSTEMATIC PROBLEMS IN THE NORTHEAST ... 12
  - SYSTEMATIC PROBLEMS IN THE SOUTHEAST ... 24
- SYSTEMATIC DESCRIPTIONS ... 66
  - Genus *Canis* Linnaeus ... 66
  - *Canis cedazoensis* Mooser and Dalquest ... 68
  - *Canis lepophagus* Johnston ... 68
  - *Canis latrans* Say ... 73
  - *Canis edwardii* Gazin ... 82
  - *Canis rufus* Audubon and Bachman ... 85
  - *Canis armbrusteri* Gidley ... 90
  - *Canis lupus* Linnaeus ... 93
  - *Canis familiaris* Linnaeus ... 102
  - *Canis dirus* Leidy ... 106
- SUMMARY ... 118
- LITERATURE CITED ... 121
- ADDENDUM ... 136
- APPENDIX A ... 138
- APPENDIX B ... 144
- APPENDIX C ... 150

INTRODUCTION

According to the revisionary work of Anderson (1943), Goldman (1937, 1944), and Jackson (1951), three living species of wild *Canis* occur in North America: *Canis latrans*, the coyote; *C. rufus*, the red wolf; and *C. lupus*, the gray wolf. Although this taxonomic arrangement has been generally accepted, some questions have arisen concerning matters not fully explained by the revisions, and certain newly recognized phenomena within canid populations.

Much attention, especially since 1960, has been directed toward the systematics of *Canis* in the eastern half of North America. There has been controversy regarding the taxonomic status of the wolves (*C. lupus lycaon* and subspecies of *C. rufus*) originally found there, and of the populations of *Canis* presently inhabiting the region (roughly, east of 100°W and south of 50°N). The production of fully fertile hybrids, of common occurrence among captive *Canis* (Gray, 1972), has been suggested as having affected wild *Canis* in the eastern part of the continent.

In addition to the questions concerning living *Canis*, there are problems involving the paleontological history of the genus in North America. Although many fossil specimens have been described, not all of them have been assigned to particular lineages ancestral to living populations. Of special interest, because it is the only fossil kind represented by what a modern mammalogist would call a good series, is the dire wolf, *Canis dirus*.

My aim in studying *Canis* was to examine large series of specimens from throughout North America, in order to obtain a clearer understanding of the systematic relationships between the species represented. I hoped to get an idea of the extent of variation within the Recent wolves found in the western and northern parts of the continent, and to determine the relative positions of *C. dirus*, *C. rufus*, and *C. lupus lycaon*. I wanted also to ascertain, as well as possible, the origins and relationships of the presently existing populations of *Canis* in the east. Partly from study of populations, I hoped to distinguish and more accurately delineate (morphologically, geologically, geographically) the living and extinct species of North American *Canis*.

Because of their recognized taxonomic value, abundance in museum collections, paleontological preservation, and relative ease of handling, I used skulls as the primary material of my study. Approximately 5,000 specimens were examined.

The first main part of the paper consists of an historical sketch and a statistical analysis based on those populations represented by large series of complete skulls. This analysis serves to delineate special groups and to assess the probable origin and relationship of questionable populations. The BMD07M program of multivariate analysis was a primary method employed in this study. The second main part of the paper consists of descriptions of each recognized species of North American *Canis*. Some of the specimens discussed, including many of the fossils, could not be used in multivariate analysis, but the descriptions are supported in part by univariate and bivariate statistics.

Collections cited in this paper are represented by the following abbreviations: AMNH, American Museum of Natural History; ANSP, Academy of Natural Sciences, Philadelphia; CM, Carnegie Museum; CNM, National Museum of Canada; FGS, Florida Geological Survey; FM, Field Museum; ISM, Illinois State Museum; KU, University of Kansas Museum of Natural History; LACM, Los Angeles County Museum of Natural History; LPI, Louisiana Polytechnic Institute Department of Zoology; LSUMZ, Louisiana State University Museum of Zoology; MCZ, Harvard University Museum of Comparative Zoology; MSU, Michigan State University Museum; NYEC, New York Department of Environmental Conservation; PPM, Panhandle Plains Museum; PUWL, Purdue University Wildlife Laboratory; QWS, Quebec Wildlife Service; ROM, Royal Ontario Museum; SD, San Diego Natural History Museum; SMUMP, Southern Methodist University Museum of Paleontology; SR, Sul Ross State University Department of Biology; TM, Texas Memorial Museum; UAlb, University of Alberta Department of Zoology; UAriz, University of Arizona Department of Biological Sciences and Laboratory of Paleontology; UArk, University of Arkansas Department of Zoology; UCMP, University of California Museum of Paleontology; UCMVZ, University of California Museum of Vertebrate Zoology; UColo, University of Colorado Museum; UF, University of Florida State Museum; UI, University of Illinois Museum of Natural History; UMMP, University of Michigan Museum of Paleontology; UMMZ, University of Michigan Museum of Zoology; UMinn, University of Minnesota Museum of Natural History; UN, University of Nebraska State Museum; UO, University of Oklahoma Museum; USFWS, United States Fish and Wildlife Service field collections; USNM, United States National Museum of Natural History; VFG, Vermont Fish and Game Department. A few other collections are spelled out in the text. Other common abbreviations in this paper include "C." for *Canis* and "A." for *Aenocyon*.

This paper is a slightly modified version of a Ph.D. dissertation (Nowak, 1973) submitted to the University of Kansas in 1973. Since that year substantial new information has become available, as for example through Kurten's (1974) study of fossil coyotes, Kolenosky and Standfield's (1975) analysis of wolves in Ontario, and Mooser and Dalquest's (1975) description of a new species of North American Pleistocene *Canis*. In addition, a number of specimens have been collected recently in southeastern Texas, and this material allows an updating of the status of the red wolf in that area. Although I have devoted some space to the newly available views and data, they have not, in all cases, received the same degree of attention shown the earlier material. Other differences between my dissertation and this paper include the dropping in the latter of several figures and tables of measurements, the relegation of the statistical analysis of the dire wolf to the section entitled "Systematic Descriptions," and the correction of several errors. I have not cited my dissertation as a reference for this paper, except in a few instances in which mention of the contrast between the two seemed warranted.

Acknowledgements

It is impossible to express my full appreciation to all who assisted me. I must, however, single out Professor E. Raymond Hall, Museum of Natural History, University of Kansas, who was my major advisor until his retirement in May 1972, and who then voluntarily continued to act in this capacity. Professor Hall initially suggested that I do graduate work at the University of Kansas, and it was his idea for me to make a study of the relationships between species of North American Quaternary *Canis*. On countless occasions he provided me with assistance and advice regarding my dissertation and my general program of work at the University. I consider it a rare honor to have been among his students.

Practically all of the other instructors, and many of the students, with whom I have been associated at the University of Kansas, have at one time or another given me some help that eventually contributed to this paper. I thank them all, but specifically want to mention Professors Robert S. Hoffmann, Robert M. Mengel, and Craig C. Black (now of the Carnegie Museum). I also am grateful to Professor Peter M. Neely, Associate Director of the University Computer Center, who took much of his time to explain the processes and results of the BMD07M computer program. Although I could not have effectively used

BMD07M without Professor Neely's help, I take full responsibility for the application and interpretation of this program with regard to the problems of my study. Additional valuable assistance on the use and understanding of computers was provided by my fellow student, Alberto Cadena.

The research required in preparation of this paper necessitated considerable travel to museums and other localities throughout North America. I therefore am especially grateful to those organizations that aided me in this regard. The Theodore Roosevelt Memorial Fund of the American Museum of Natural History, and the National Science Foundation each made a direct grant for travel and related expenses. The Committee on Systematics and Evolutionary Biology, University of Kansas, provided travel grants in 1970 and 1971, a research assistantship in the summer of 1972, and also a traineeship for the academic year 1971-1972.

No progress could have been made in my research had it not been for the cooperation of numerous persons who generously assisted me in the examination of specimens and associated materials in their care. I want to especially thank John L. Paradiso, Bird and Mammal Laboratories, United States National Museum of Natural History (now of the Office of Endangered Species, U.S. Fish and Wildlife Service). During the four months that my wife and I worked at the National Museum, he aided us in every way possible and spent a great deal of his own time to see that we were well provided for both in and out of the Museum. John and I actually have been in close communication regarding *Canis* since 1965. Many of the views expressed in this paper were developed jointly with him in the course of years of pleasant study, conversation, and correspondence.

I am also grateful to the following persons who either sent me specimens on loan or assisted me when I visited their areas: Sydney Anderson, American Museum of Natural History; Rollin H. Baker, The Museum, Michigan State University; Troy L. Best, Museum of Zoology, University of Oklahoma; Elmer C. Birney, Museum of Natural History, University of Minnesota; Ben Day, Vermont Fish and Game Department; Diana Van Elsacker, University of Colorado Museum; David E. Fortsch, Los Angeles County Museum of Natural History; Philip S. Gipson, Department of Zoology, University of Arkansas; John W. Goertz, Department of Zoology, Louisiana Polytechnic Institute; John E. Guilday, Carnegie Museum; C. R. Harington, National Museum of Canada; Billy R. Harrison, Panhandle Plains Museum; Claude W. Hibbard, Museum of Paleontology, University of Michigan; Donald F. Hoffmeister, Museum of Natural History, University of Illinois; Emmet T. Hooper, Museum of Zoology, University of Michigan; J. H. Hutchison, Museum of Paleontology, University of California; Frederick F. Knowlton, U.S. Fish and Wildlife Service; Barbara Lawrence, Museum of Comparative Zoology, Harvard University; Everett H. Lindsay, Laboratory of Paleontology, University of Arizona; George H. Lowery, Jr., Museum of Zoology, Louisiana State University; Ernest L. Lundelius, Jr., Texas Memorial Museum; Larry D. Martin, Museum of Natural History, University of Kansas; John D. Newsom, Louisiana State University Cooperative Wildlife Research Unit; Robert T. Orr, California Academy of Sciences; N. Panter, Department of Zoology, University of Alberta; Paul W. Parmalee, Illinois State Museum; Oliver P. Pearson, Museum of Vertebrate Zoology, University of California; Randolph L. Peterson, Royal Ontario Museum; Charles Pichette, Quebec Wildlife Service; Douglas H. Pimlott, University of Toronto; Clayton E. Ray, U.S. National Museum of Natural History; Richard L. Reynolds, Los Angeles County Museum of Natural History; Horace G. Richards, Academy of Natural Sciences, Philadelphia; Glynn Riley, Jr., U.S. Fish and Wildlife Service; C. B. Robbins, Department of Biological Sciences, University of Arizona; Dennis N. Russell, Texas Parks and Wildlife Department; Donald Schierbaum, New York Department of Environmental Conservation; James F. Scudday, Department of Biology, Sul Ross State University; Beryl E. Taylor and Richard H. Tedford, American Museum of Natural History; Gilmer Voss, San Diego Natural History Museum; S. David Webb, University of Florida State Museum; J. William Yon, Florida Geological Survey; Phillip M. Youngman, National Museum of Canada; and Curtis J. Carley, U.S. Fish and Wildlife Service.

These acknowledgements would not be complete without the names of my parents, Jacob and Esther Nowak, New Orleans, Louisiana. Throughout the course of my research they were always ready and willing to provide any assistance, whether requested or not.

I finally wish to express my gratitude to my wife, Thu. Although having only recently arrived in the United States, and with an incomplete command of the English language, she served as an indispensable full time assistant, especially in the recording of data. Subsequently, she prepared the base maps and parts of the figures herein.

Methods

As a primary statistical tool I employed the Biomedical computer program, number 07M, stepwise discriminant analysis (Dixon, 1970). This method is a modified version of multivariate discriminant function analysis, as used previously in the study of *Canis* and explained in detail by Jolicoeur (1959), Giles (1960), Lawrence and Bossert (1967), and Gipson (1972). The BMD07M program involves a procedure known as canonical analysis, as discussed by Rao (1952) and Seal (1964).

In multivariate analysis a series of variables from an individual specimen are considered together to determine the position of that specimen relative to other specimens. In its simplest form this procedure resembles that of a scatter diagram in which the location of a specimen on a two dimensional graph is determined by its position along both a vertical and horizontal axis, each representing a single variable. The multivariate analysis, through a process of matrix inversion, can consider numerous variables, but plots the results in the same form of a two dimensional graph.

The BMD07M program requires that at least two designated groups of individuals be entered into the analysis. The variables are tested one at a time for their ability to distinguish between the groups. If any variable is found to have too low a discriminatory power, that variable is rejected and not considered in the analysis. The effects of correlation among the variables are eliminated in this program by a process of eigenvalue extraction.

On the basis of the variables selected, the designated groups are separated as well as is possible. The statistical distance between groups (D^2 of Mahalonobis), calculated from the combined variables, may be printed out if desired. In addition, each individual specimen is given a D^2 distance from each group, and is assigned canonical coordinates to plot its position relative to all other specimens. If the variables employed have effectively distinguished the groups, the specimens within a particular group will be nearer to each other than to the specimens of other groups. Once definite groups have been established, specimens of questionable identity may be individually entered into the analysis to determine their position relative to the groups and hence their possible taxonomic affinity.

For use in multivariate analysis, the 15 measurements listed in appendix B were selected. These measurements were considered to represent all of the main dimensions of the skull plus those of three of the more diagnostic teeth. Additional measurements, especially of the teeth, which are individually of diagnostic value, could have been added.

But large series of specimens were desirable, and so measurements of parts too often missing or defective were excluded. Also it was reasoned that the 15 utilized measurements would adequately express the major functions of the skull. Because the entire mandible was occasionally missing from specimens, measurements of the lower jaw and teeth were omitted in multivariate analysis. Several test runs of the program, involving as many as 35 measurements, including those of the mandible, did not seem to produce results different from those that follow, nor to noticeably increase the discriminatory ability of the analysis.

Lawrence and Bossert (1967) divided each of 15 measurements by greatest length of skull and entered their analysis with the resulting series of fractions, intending thereby to eliminate size as a discriminating factor. Actually there are various expressions of size of a skull, and dividing by any one of them may produce different results. Furthermore, it is questionable whether any attempt should be made to eliminate the size factor, because it appears to be a definite biological factor, at least in distinguishing the wild species of North American *Canis*. Certain skulls, representing two kinds of *Canis* that would not ordinarily be confused because of size differences, may have similar proportions of greatest length to other measurements.

Therefore raw measurements were used in most of the following calculations. This procedure considers the size of each measurement simultaneously as a factor in classifying a specimen. Since the sizes of the various measurements may vary at different rates between different species, proportion is also a factor in the analysis. It is true that an unusually large or small specimen may be assigned to a group other than that which its proportions indicate, but such occurrences are rare. In any case, my tests of this particular computer program, using the selected 15 measurements, revealed that in most instances variables based on raw measurements and fractions of greatest length of skull produced similar depictions of relationship, but that the raw measurements gave a wider separation between groups.

With one major exception (dogs, see below), different analyses were used for males and females, and it was found that such a procedure usually produced wider separation between groups than was achieved by combining sexes. This wider separation occurred regardless of whether raw measurements or fractions of greatest length were used as variables.

The sex of some of the skulls utilized in statistical analysis was unknown, and these skulls were assigned to male or female categories on the basis of size and the other factors explained below. Fortunately, excepting domestic dogs, each of the major standard groups, against which other material was tested, consisted predominantly of specimens of known sex. In the subsequent pages, when a sample size of one sex is listed, it is followed by the number (in parentheses) of specimens in the series (if any) for which sex had not been recorded, but which were judged to belong to that sex.

In the statistical analyses, specimens of domestic dogs (*C. familiaris*) were not separated by sex. Dog skulls are poorly represented in museum collections, compared with skulls of wild *Canis*, and less than half of the 50 specimens of *C. familiaris* used in my analyses were of known sex. Individual variation in this species is so great that it tends to obscure sexual differences in the morphology of the skull. Consequently, dogs of male, female, and unknown sex were combined in one group.

In addition to multivariate analyses, tables of measurements, with means, extremes, standard deviations, and coefficients of variation, are provided in appendix B. In some cases I also have drawn ratio diagrams that depict differences in size and proportion between the specimens of various groups. Whereas multivariate analysis demonstrates the collective results of such differences, the ratio diagram permits visualization of how each group differs in individual measurements. Simpson (1941) explained this method in detail in the course of his account of Pleistocene felines. Briefly, raw statistics (individual measurements, means of a series, etc.), taken on two or more specimens or series, are converted to their logarithms. One of the specimens or series is taken as a standard, and the difference is found between the logs of its individual measurements or means and the respective logs of the other specimens or series. In diagramming, the standard values are all plotted in a vertical line representing the zero point, and the respective values of the other specimens or series are plotted at a horizontal distance from the standard values, representing the difference between the two values.

Many skulls were examined for which the complete set of 15 measurements, required in multivariate analysis, could not be obtained because of damage, wear, or missing parts. Except for fossil material, data from such specimens were not incorporated in the ratio diagrams or statistical tables (appendix B). Therefore, the groups represented in the multivariate analyses, ratio diagrams, and tables are all of identical composition. Specimens not used for the calculation of statistics did not appear to differ from the main series.

I finally want to make it clear that I used multivariate analysis in a supporting and demonstrative role, rather than as a problem solver in itself. The analysis did not provide any major conclusions that were not apparent from more conventional methods of examination, but it did allow the efficient evaluation of many data, and the objective, graphical portrayal of a complex situation.

Age and Secondary Sexual Variation

The aging process in *Canis* was described by Goldman (1944:400-401), Jackson (1951: 250-251), Miller, Christensen and Evans (1965:652-653), and Mech (1970:139-143). Gier (1968:54-55) showed how to estimate the age of *C. latrans* by examination of wear on the incisor and canine teeth. Linhart and Knowlton (1967) demonstrated a method of aging coyotes through evaluation of cementum layers in the canine teeth.

By the age of six months in *Canis*, the permanent dentition, except for the canine teeth, is fully in place, and the skull has reached approximately 90 percent of its eventual total length. Complete emergence of the canines, and maximum dimensions of the skull, however, are not attained until about 12 months in coyotes and 15 months in larger gray wolves. Therefore, for the calculation of statistics in the following sections of this paper, I used only skulls of animals estimated to be at least 12 months old, and did not use some wolves that were under 15 months old.

The males of *Canis* average larger than the females in every measurable dimension of the skull, but there is extensive overlap between the two sexes. Males have proportionally broader rostra and higher sagittal crests. In many female coyotes the sagittal crest is flattened, and the temporal ridges that usually coalesce in males are in some females lyrate.

Statistical comparison was made of measurements of skulls of 97 male and 61 female *C. latrans lestes* from Colorado and Idaho, and of skulls of 51 male and 35 female *C. lupus mogollonensis, youngi* and *irremotus* from the mountainous region of the western United States. In this particular test, only specimens of known sex were used. Each species was examined separately and a large overlap of the two sexes was found in all 15 of the measurements considered. Males averaged larger in each measurement, however, and analysis of variance and STP tests showed a significant difference (p less than .05) between the males and females of each species in all measurements except postorbital constriction of braincase.

HISTORY AND STATISTICAL ANALYSIS OF RECENT POPULATIONS

Comparison of Known Series of Dogs, Wolves, and Coyotes

The questions to be considered in this paper concern primarily eastern North America and fossil history. Is the red wolf of the southeastern United States a full species, a subspecies of the gray wolf, a subspecies of the coyote, or a hybrid between *C. lupus* and *C. latrans*? What is the origin and affinity of the populations of *Canis* that recently have become established in much of the eastern half of the continent? Is hybridization a major factor in the situation? Is the Pleistocene dire wolf completely distinguishable from the modern gray wolf, and is it possibly ancestral to the latter?

Before attempting to answer these and other questions, it would be advisable to delineate the perimeters of those populations that seem best to represent recognizable species. Throughout most of that part of North America in which the coyote, gray wolf, and domestic dog are found together, they are easily distinguishable and usually behave toward one another as species. The gray wolf once occurred in all of North America except for parts of the southeastern United States, most of the state of California, Baja California and the coastal lowlands of Mexico, and the region south of central Mexico (Goldman, 1944:414). A record of *C. lupus baileyi* from Tequisistlan, Oaxaca, southern Mexico (Goodwin, 1969:224) seems to have been based on questionable evidence. The coyote was originally found throughout most of the western half of the continent, and its range in the northeast extended as far as the upper Great Lakes (Young, 1951:29). The domestic dog, *C. familiaris*, has long occurred in all parts of the continent, almost always in association with man. All dogs may have descended from a small southwest Asian subspecies of *C. lupus* that was domesticated 10 to 12 thousand years ago (Scott, 1968). The dogs of the American Indians were apparently introduced into the New World by man, and do not seem to have been influenced by interbreeding with native species of North American *Canis* (Allen, 1920; Haag, 1948). Specimens of the earliest known domestic dogs on the continent were described from a site in Lemhi County, Idaho dated at 10,400-11,500 B.P. They reportedly already possess the typical characters of *C. familiaris* (Lawrence, 1966, 1968).

Only in the eastern part of North America do hybridization and modification of the original populations appear to be of possible significance. A few isolated instances in which *C. familiaris* hybridized with either *C. lupus* or *C. latrans* in other regions, have been reported (Young, 1944:180-210; Mengel, 1971; Gray, 1972), but such cases do not seem to have had lasting effect on populations. No instances of interbreeding between *C. lupus* and *C. latrans* in the western half of the continent have yet been reported. Therefore, it is reasonable to consider *C. lupus* and *C. latrans* of the western and northern parts of the continent as consisting of natural, unmodified populations that may confidently be used as a basis on which to test more questionable populations.

For an initial analysis of known groups, I decided upon using skulls of *C. latrans* and *C. lupus* that had been collected not later than 1925 in the mountainous region of the west. The gray wolf sample included 57(6) males and 37(2) females (parentheses contain numbers of specimens in the series for which sex had not been recorded, but which were judged to belong to the particular sex indicated; see p. 5). This group consisted of all skulls of adult *C. lupus mogollonensis, youngi,* and *irremotus* in the U.S. National Museum of Natural History, upon which the 15 necessary measurements could be made, except for two specimens taken after 1925 (see appendix A, part 1).

Of coyotes, 97 male and 61 female skulls, all of the subspecies *C. latrans lestes,* were utilized in the initial test (see appendix A, part 2). These specimens comprised the entire National Museum collection of Idaho and Colorado adult *lestes* of known sex, taken prior to 1926, except for skulls upon which all of the needed 15 measurements could not be made.

The selection of these particular specimens as standard comparative material had the following advantages: (1) the wolf and coyote had long been sympatric in the region, and thus theoretically would have evolved the maximum amount of differential characters reflecting their separate ecological niches; (2) the region had a minimum human (and presumably domestic dog) population; (3) the time period was one in which both the wolf and coyote were common (most of the specimens represent the first few years of Federal predator control work which began in 1915); and (4) the subspecies of both *C. lupus* and *C. latrans* do not exhibit extremes of size or other characters within their respective species.

For a sample of domestic dogs, only those skulls were selected which, while known to be *C. familiaris,* were superficially nearest to those of *C. lupus* or *C. latrans* in appearance. The extremes of domestication represented by broad-skulled dogs (as bulldogs), narrow-skulled dogs (as Russian wolfhounds), and dogs having greatly reduced rostra (as pugs) were avoided. Extremely small dogs, those in which the skull was less than 150 millimeters in greatest length, also were not used. Specimens utilized in the sample included 1 Eskimo dog, 5 Irish wolfhounds, 3 German shepherds, 2 sheep dogs, 2 Newfoundlands, 1 doberman pinscher, 1 greyhound, 1 great Dane, 1 mastiff, 1 Irish setter, 1 beagle, 1

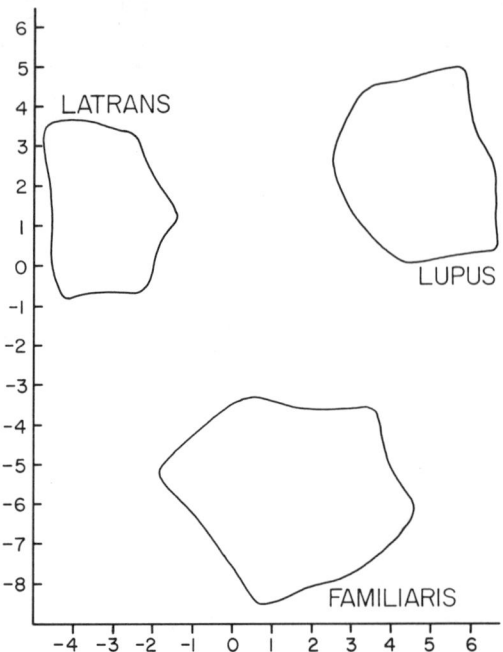

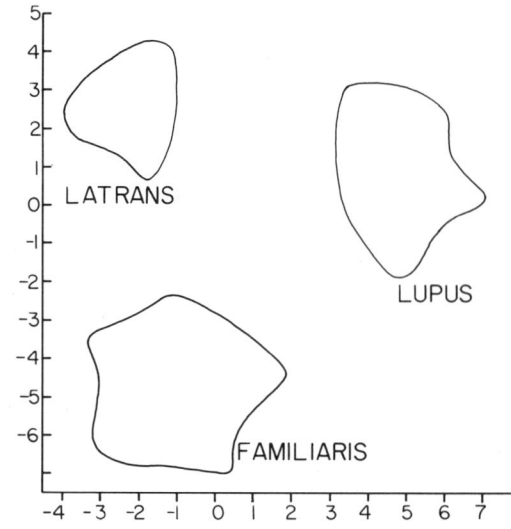

Fig. 1.—Graphical results of multivariate analyses comparing samples of *C. lupus* and *C. latrans* from the mountainous region of western North America, and *C. familiaris.* Only the margins of the range of variation of each species are shown. In this and in all subsequent portrayals of multivariate analyses, the numbers along the vertical and horizontal axes are canonical coordinates. These coordinates are used to indicate relative position, and do not represent any material values. In this figure and subsequent portrayals of analyses, males are shown above and females are shown below.

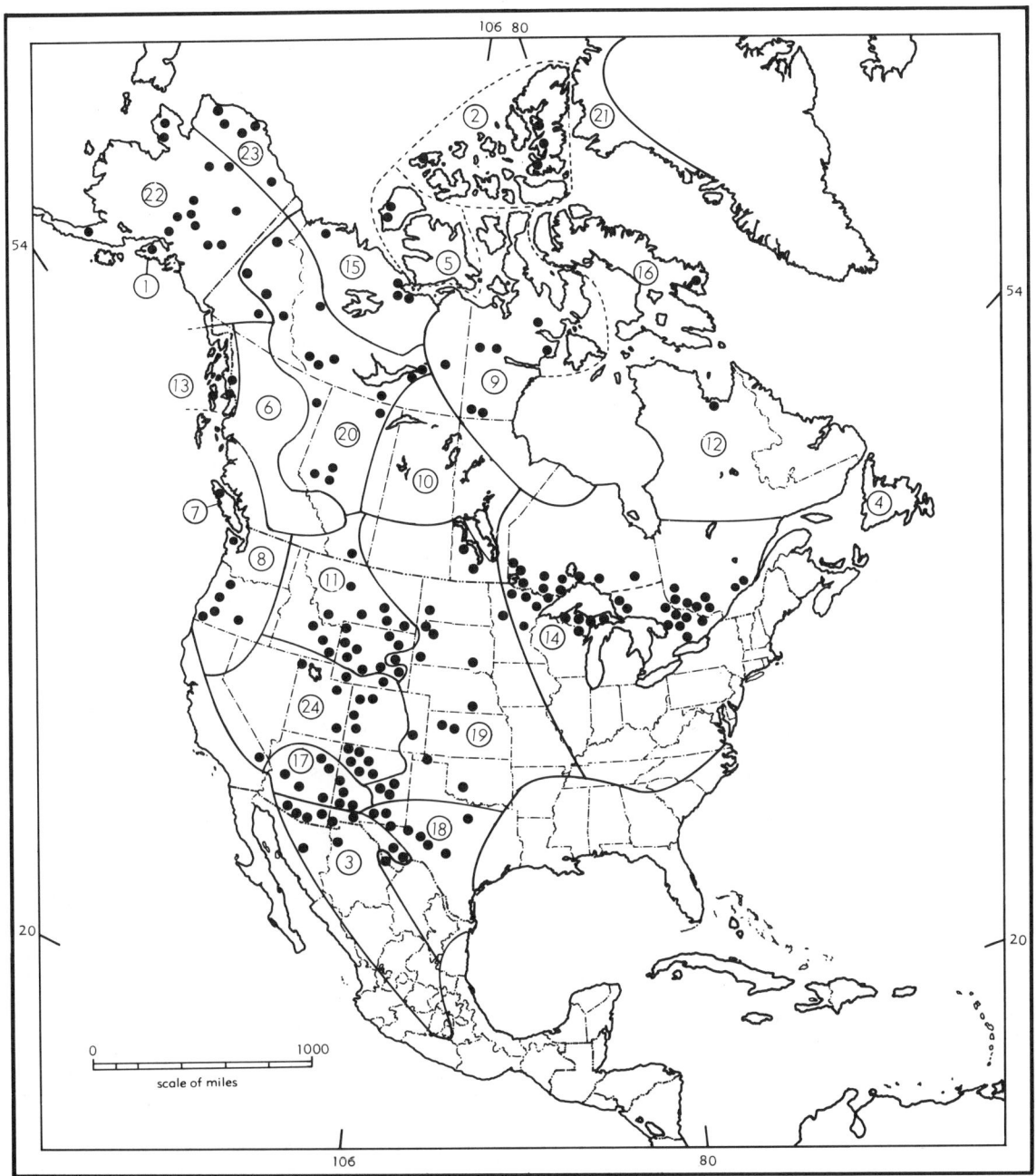

Fig. 2.—Range map of *C. lupus* in North America showing localities (black dots) of specimens used in the statistical analyses of this paper. The numbers on the map represent recognized subspecies, as follows:

1. *C. l. alces*	7. *C. l. crassodon*	13. *C. l. ligoni*	19. *C. l. nubilus*
2. *C. l. arctos*	8. *C. l. fuscus*	14. *C. l. lycaon*	20. *C. l. occidentalis*
3. *C. l. baileyi*	9. *C. l. hudsonicus*	15. *C. l. mackenzii*	21. *C. l. orion*
4. *C. l. beothucus*	10. *C. l. griseoalbus*	16. *C. l. manningi*	22. *C. l. pambasileus*
5. *C. l. bernardi*	11. *C. l. irremotus*	17. *C. l. mogollonensis*	23. *C. l. tundrarum*
6. *C. l. columbianus*	12. *C. l. labradorius*	18. *C. l. monstrabilis*	24. *C. l. youngi*

The solid lines indicate subspecific boundaries. The dashed line in southeastern Ontario shows Standfield's (1970) division between his "Ontario type" and "Algonquin type" of *C. lupus lycaon*. Because of the scale of the map, it was not possible to plot all localities in crowded areas.

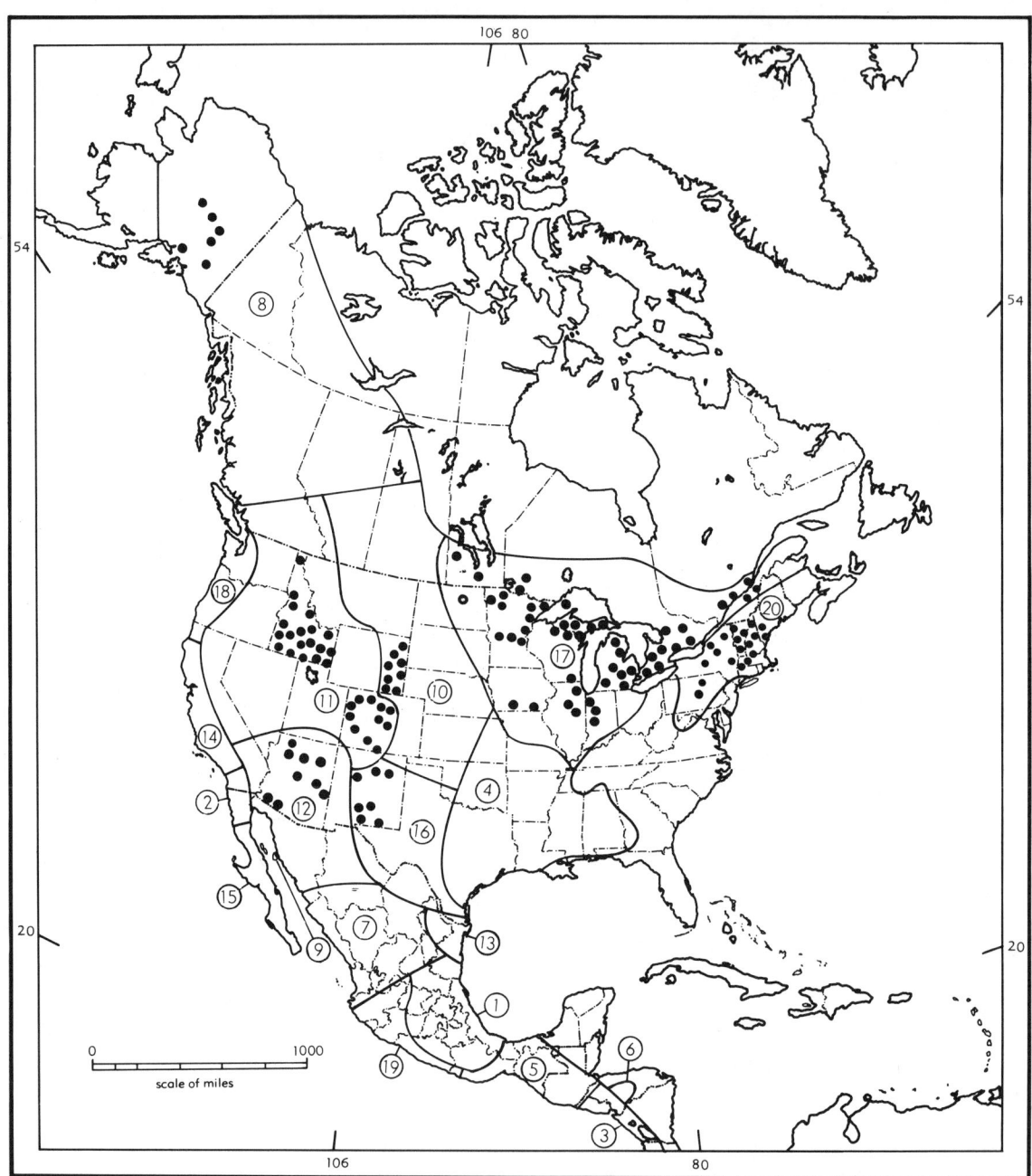

Fig. 3.—Range map of *C. latrans* showing localities of specimens used in the statistical analyses of this paper. The numbers on the map represent recognized subspecies, as follows:

1. *C. l. cagottis*
2. *C. l. clepticus*
3. *C. l. dickeyi*
4. *C. l. frustror*
5. *C. l. goldmani*
6. *C. l. hondurensis*
7. *C. l. impavidus*
8. *C. l. incolatus*
9. *C. l. jamesi*
10. *C. l. latrans*
11. *C. l. lestes*
12. *C. l. mearnsi*
13. *C. l. microdon*
14. *C. l. ochropus*
15. *C. l. peninsulae*
16. *C. l. texensis*
17. *C. l. thamnos*
18. *C. l. umpquensis*
19. *C. l. vigilis*
20. *C. l.* "var." (Lawrence and Bossert, 1969)

The solid lines indicate subspecies boundaries. Because of the scale of the map, it was not possible to plot all localities in crowded areas. Certain additional localities of *C. latrans* are shown in Figs. 14, 25, and 31.

basset hound, and 30 dogs of unknown or mixed breed (see appendix A, part 3). Of these 50 specimens, only 11 males and 9 females had been previously identified as to sex. For the reasons explained above, domestic dogs of male, female, and unknown sex were combined into a single sample.

The graphical results of multivariate analyses of males and females of the initial samples are depicted in figure 1. For both sexes there is complete separation between all three species. Such an arrangement could be expected, and it may serve as a sound basis on which to evaluate other specimens.

The next step was to compare skulls taken elsewhere in northern and western North America to the above series of reliably distinguished specimens. Skulls of 176(27) males and 114(33) females, previously identified as *C. lupus*, and of 69(2) males and 50(4) females, identified as *C. latrans*, were tested individually against the three known groups (see appendix A, parts 4 and 5). The maps in figures 2 and 3 show localities of all specimens of gray wolves and coyotes. Figure 4 shows the results of multivariate analyses. Nearly all of the newly added material falls within the range of variation of the appropriate original sample, or at least is closer to this range than to that of other species.

Only five of these specimens (see appendix A, part 6) seem confusing as to identity. Three skulls, previously identified as *C. lupus baileyi*, are statistically and morphologically intermediate to known samples of female gray wolves and coyotes. *Canis lupus baileyi*, the smallest subspecies of North American gray wolf, shared its entire range with *C. latrans*, and interbreeding between the two might have been possible under certain conditions. I thus consider these three specimens as probable hybrids of *C. lupus* and *C. latrans*, and henceforth have not used them in the formation of samples of either parent species.

Two females from the Sacramento Mountains of New Mexico, previously identified as *C. lupus monstrabilis*, appear both visually

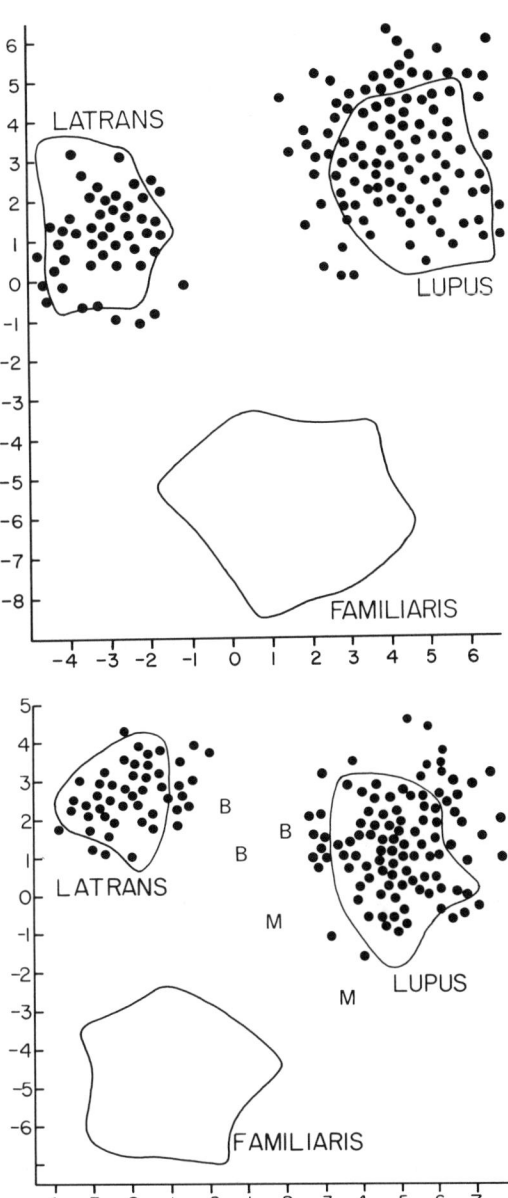

Fig. 4.—Statistical positions of individual native, wild-caught specimens from throughout northern and western North America, relative to ranges of variation of the series of *C. lupus*, *C. latrans*, and *C. familiaris* shown in Fig. 1. The black dots represent individuals of *C. lupus* and *C. latrans*, all of which fall close to the appropriate range of variation. The letter B represents specimens throught to be hybrids between *C. lupus baileyi* and *C. latrans*; the letter M represents specimens thought to be hybrids between *C. lupus monstrabilis* and *C. familiaris*. Males are above, females below; numbers along axes are canonical coordinates.

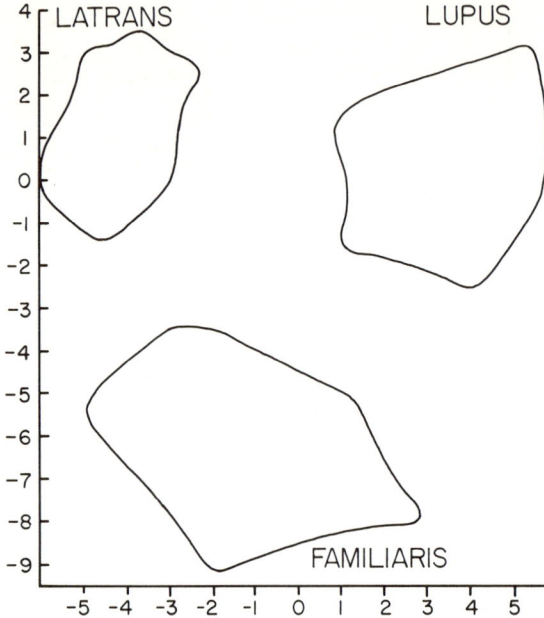

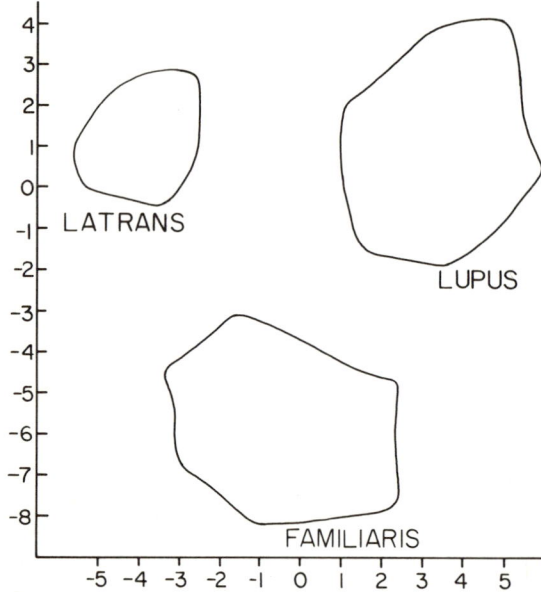

Fig. 5.—Multivariate comparison of all specimens of *C. lupus* and *C. latrans* from northern and western North America, and *C. familiaris*. Only the margins of the range of variation of each species are shown. Males are above, females below.

and statistically (Fig. 4) to be intermediate to the gray wolf and domestic dog. At least two other skulls collected in New Mexico, but not complete enough for inclusion in the multivariate analysis, also suggest the occurrence of hybridization between the two species. Nonetheless, the preponderance of material that can be clearly identified as either *C. latrans* or *C. lupus*, indicates that in those regions hitherto discussed the relationships of these two canids to one another and to *C. familiaris* were those of normal species.

All of the northern and western specimens, except the five considered to be hybrids, were incorporated with the appropriate standard samples of *C. lupus* or *C. latrans*, and these two groups along with the sample of 50 domestic dogs were tested in new multivariate analyses. The graphical results depicted in figure 5 once again indicate clear separation between the three species. Measurements for the total series of dogs, and of western and northern wolves and coyotes are listed in appendix B (parts 1 and 2). The means of these measurements (of males only for *C. lupus* and *C. latrans*) are compared in the ratio diagram in figure 6.

Systematic Problems in the Northeast

Decline of the Gray Wolf

According to Goldman (1944), a single subspecies of gray wolf, *C. lupus lycaon*, originally occupied the region from eastern Minnesota to the Atlantic, and from northern Ontario to parts of the southeastern United States. Considering the enormity of this range, however, and the problems associated with the systematics of *Canis* in eastern North America, Goldman used relatively few specimens for describing the situation. From the entire region south of Lakes Michigan and Erie he assigned to *lycaon* only four complete skulls and one mandibular ramus.

Standfield (1970) reported the presence of two distinct kinds of *lycaon* in Ontario, which he designated the "Ontario type" and the "Algonquin type." The former was said to occur mainly in the boreal forests north

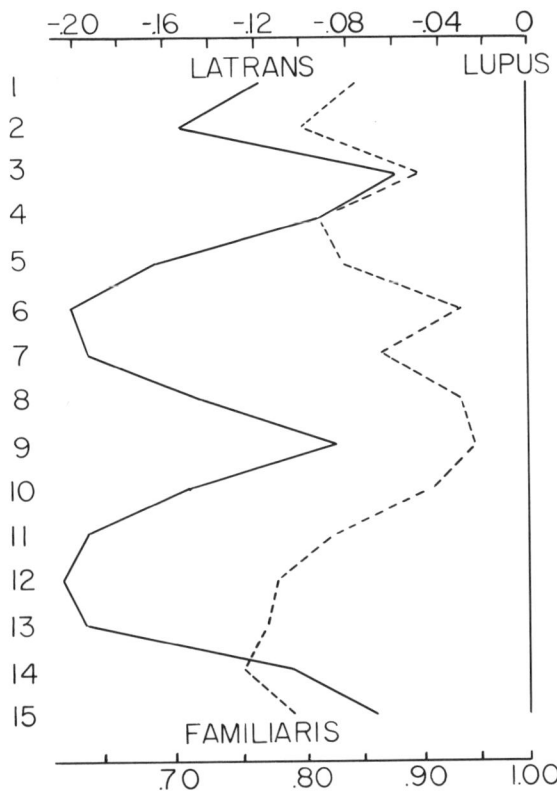

Fig. 6.—Ratio diagram comparing means of total samples of *C. lupus* and *C. latrans* from northern and western North America (males only are shown), and *C. familiaris* (dashed line). Vertically arranged numbers represent the measurements so numbered in Appendix B. A log difference scale is provided above, and a ratio scale below the diagram.

and northeast of Lake Superior, to be larger, and to vary in pelage from pure white to jet black. The latter was said to occur in the deciduous forests east and southeast of Lakes Superior and Huron, to be smaller, and to be invariably gray-brown. Kolenosky and Standfield (1975) made a multiple discriminant analysis comparing 105 skulls of the "Algonquin type" and 122 skulls of the "Ontario type" (now referred to as the "Boreal type"), and concluded that there were significant differences between the two. These authors (p. 71) reported that "the ranges of the two types overlap throughout a broad band across eastcentral Ontario, but there is no conclusive evidence of their interbreeding."

Mech and Frenzel (1971) suggested that the present population of wolves in northeastern Minnesota consists at least in part of the subspecies *C. lupus nubilus*. This idea was based on their observations of black wolves and white wolves in the area. Individuals of these colors were common among *nubilus*, but were not reported in cited observations of *lycaon* in eastern Ontario.

Whatever their original systematic status, northeastern wolves have suffered a drastic loss in numbers and range because of persecution by Caucasian man. The species practically disappeared south of the St. Lawrence River between 1850 and 1900 (Peterson, 1966:200; Goodwin, 1936), and was gone from southern Michigan and Wisconsin early in the twentieth century (Jackson, 1961:293; Arnold, 1952). Wolves were reported to be present in moderate numbers in the upper peninsula of Michigan by Stebler (1944), and in northern Wisconsin by Schorger (1942). In the 1950's, however, the wolf populations of these two states declined sharply. Jackson (1961:293) estimated 50 wolves to be present in northern Wisconsin, and Keener (1970) reported that none still survived in the area. For the upper peninsula of Michigan, recent numerical estimates have been 20 (Smits, 1963), less than 12 (Douglass, 1970), and about six (Hendrickson and Robinson, 1975). In March 1974 four wolves from northern Minnesota were released in upper Michigan, but by September all had been killed through human agency (Weise, et al., 1975). A viable group of about 20 to 30 individuals, however, has maintained itself on Isle Royale in Lake Superior since the late 1940's (Mech, 1966; Wolfe and Allen, 1973).

The only major population of *C. lupus* to be found anywhere in the United States south of Canada is that in northern Minnesota. An estimated 1,000 to 1,200 individuals are reported to exist in the area and they are said to be in no immediate danger of extirpation (Mech, 1977).

Despite intensive control measures, wolves reportedly still occupy all of the forested parts of Ontario, even areas close to Toronto and Ottawa. They do not, however, enter the settled agricultural sections of that province (Clarke, 1970). Here and in other parts of the northeast, there evidently has been a correlation between the decline of the gray wolf and the intensity of human population and agricultural development.

Rise of the Coyote

At the time the first white settlers arrived in North America, coyotes were apparently confined to open plains and more arid regions, mainly in the western half of the continent (Young, 1951; Seton, 1929). The original range of the species did, however, follow the prairie peninsula through the midwestern states, at least as far as northeastern Indiana (Mumford, 1969:85). According to Jackson (1961:285) coyotes were undoubtedly present in southern Wisconsin when the early explorers arrived, and a few may have inhabited the northern part of the state. The name *C. latrans thamnos* was applied by Jackson (1949) to the coyote of the northeastern portion of the range of the species.

Sometime after the middle of the nineteenth century, coyotes began appearing to the north, east, and south of the prairies, and by the mid-twentieth century they existed in large numbers beyond their original range (De Vos, 1964; Mech, 1959, 1961; Young, 1951). Many coyotes had escaped from captivity, or had been deliberately released by sport or bounty hunters. Some of these animals formed local breeding populations that maintained themselves over a period of time. Records that seem attributable to such introductions are as follows.

Florida.—Palm Beach County, 1925; Collier, Monroe, and Marion counties (Young, 1951:15); DeSoto County, 1933 (Sherman, 1937); Polk County, 1962 (Cunningham and Dunford, 1970).

Alabama.—Barbour County, 1924-1929 (Young, 1951:16); Bazemore, Fayette County, 1956; Huntsville Arsenal, Madison County, 1961 (Holliman, 1963).

Georgia.—North Georgia, 1929; Habersham County, 1930; Ware County (Young, 1951:15).

South Carolina.—Aiken County; Edisto Island, Colleton County, 1924 (Golley, 1966: 141).

North Carolina.—Gaston County, 1938 (Young, 1951:14); Swain County, 1947 (Linzey and Linzey, 1968).

Tennessee.—Grand Junction, Hardeman County; Maury County, 1930; McCains, Maury County, 1931 (Kellogg, 1939:267); Hickman and Maury counties, early 1930's (Young, 1951:15); Benton, Hickman, and Sequatchie counties (Schultz, *et al.*, 1954:205); Tennessee National Wildlife Refuge, Henry County, 1951 (Schultz, 1955).

Kentucky.—Near Fayette-Clark county line, 1953 (Gale and Pierce, 1954).

Virginia.—Rockingham, Highland, and Grayson counties (Handley and Patton, 1947: 140); Tazewell and Lee counties, 1968-1969; near Mossy Creek, Augusta County, 1970 (Carpenter, 1971).

West Virginia.—Tucker County (Handlan, 1946).

Maryland.—5 mi. NW Poolsville, Montgomery County, 1921 (Jackson, 1922); outskirts of Baltimore, 1931 (Redington, 1931: 27); Cecil County, 1961 (Paradiso, 1969: 134).

New Jersey.—Near Ringoes, Hunterdon County, 1938 (Young, 1951:16); near Fishing Creek, Cape May County, 1948 (Ulmer, 1949).

Pennsylvania.—Clinton County, 1915-1916 (Shoemaker, 1917:11); near Flowing Spring, Blair County, 1907 (Gifford and Whitebread, 1951:46); Sheshequin Township, Bradford County, 1939; Chestnut Hill section, Philadelphia, 1942 (Ulmer, 1949); Beaver and Forest counties, 1946-1947 (Richmond and Rosland, 1949:34).

New York.—Near Ithaca, Tompkins County, 1920 (Seton, 1929:369); Ontario County, about 1928 (Bump, 1941:415).

Massachusetts.—Vicinity of Amherst, Hampshire County, 1936 (Warfel, 1937).

Maine.—Near Portland, 1932 (Aldous, 1939).

Some other occurrences in New England and New York probably represent range extensions and are discussed below. But it is difficult to determine whether some records should be designated as introductions or range extensions.

The actual eastward extension of the coyote's range has seemingly been slowest in the region between Michigan and the Ohio Valley, possibly because of the intensity of human population there. Some sources, such as Seton (1929:368) even suggest that the coyote had been exterminated in part of this region, although Mumford (1969:85) doubted that the species had ever been extirpated in Indiana. Jackson (1961:285) noted a reversal of the original situation in Wisconsin, in that coyotes had become much more common in the northern half of the state than in the southern part. Burt (1946) observed that the species was rare in the southern part of the lower peninsula of Michigan. Hoffmeister and Mohr (1957:119) stated: "The coyote occurs in much of Illinois, but it is not common anywhere in the state." And Mumford (1969:84) reported coyotes to be present throughout Indiana, but not to be common. Hamilton (1943:178) indicated that coyotes were of sporadic occurrence in most of western Ohio, and that they had been established in Logan County for 12 years. Other occurrences in the western counties of Ohio were reported by Negus (1948), Whitacre (1948), Young (1951:15), and Goodpaster and Hoffmeister (1968). Wilson (1976) mentioned the presence of a sparse coyote population in western Kentucky.

The coyote's range seems to have expanded mainly to the northeast of the original distribution in the prairie peninsula. At the base of the lower peninsula of Michigan, individuals were reported in Berrien County in 1900-1901 (Wood and Dice, 1924) and in Washtenaw County in 1905 and 1910 (Wood, 1922). A specimen was taken in Genesee County, in the east-central part of the state in 1917 (Wood and Dice, 1924), and an individual was reported in Charlevoix County, at the northern tip of the lower peninsula, in 1919-1921 (Dice, 1925).

Coyotes had also entered the upper peninsula of Michigan by the early twentieth century. Shiras (1921:166) reported: "In the past fifteen years the coyote unexpectedly appeared in northern Wisconsin and Michigan, coming from Minnesota. It has since become very numerous." Wood and Dice (1924) listed occurrences in five upper peninsula counties between 1912 and 1915.

Even earlier, coyotes had begun to move north in Minnesota. According to Bailey (1929) they first appeared in Sherburne County in 1875, and Cahn (1921) reported them to be more common than wolves in Itasca County.

Snyder (1938) thought that coyotes were present in the western Rainy River District, Ontario in 1890. Peterson (1966:197) noted that in Ontario before 1900 the species was restricted to Rainy River and western Kenora districts. Krefting (1969) suggested that coyotes moved to Isle Royale in Lake Superior sometime prior to 1912-1913, from the Sibley Peninsula area of Ontario, where they had arrived about 1900. Coyotes continued to inhabit Isle Royale until gray wolves, in a reversal of the general trend in the northeast, occupied the island in the 1940's.

From western Ontario, coyotes spread eastward above Lake Superior, and northward toward Hudson Bay. They also apparently crossed the St. Clair River into southeastern Ontario where the first specimen was taken, north of Thedford, Lambton County, in 1919. In 1943 specimens were collected in Essex, Peterborough, and Carleton counties, Ontario

(Anderson, 1946), and by 1956 the species was reported to occur nearly throughout the province (Peterson, 1957). The first record for Quebec was a specimen taken near Luskville, Gatineau County in 1944 (Rand, 1945). According to Wolfram (1964) the first coyote to be found in New Brunswick was killed near Sussex in 1958. Subsequent expansion of the species in Quebec, as far as the Gaspe Peninsula, and in New Brunswick, was reported by Georges (1976).

Bromley (1956) summarized the history of the coyote in northern New York. Individuals were shot in the St. Lawrence River area in 1925 and in Franklin County in the mid-1930's. The species was said to have achieved a good foothold in the early 1940's when its range included several areas of the Adirondack Mountains. By the late 1940's coyotes had spread throughout the Adirondacks, and in the early 1950's they occupied the entire northern third of the state. Subsequently, according to Severinghaus (1974a, 1974b), there have been records from much of southern New York. An estimate of 5,000-15,000 coyotes for the state was published by Marvinney (1976).

Coyotes apparently are continuing to move down the Appalachian Mountains from the northeast. They are currently reported to be rare, but widely distributed in Pennsylvania (John L. George, Department of Wildlife Management, Pennsylvania State University, pers. comm.), and their "sporadic presence" in West Virginia was discussed by Taylor, Counts, and Mills (1976). In May 1976 a specimen was taken near Nestorville, Barbour County in northeastern West Virginia, and I found the skull to closely resemble those of some New England coyotes.

In 1936 a coyote reportedly was killed in Argyle Township, Penobscot County, Maine. Over the next two years 11 other individuals were taken in that vicinity, most of which were considered to be hybrids between *C. latrans* and *C. familiaris* (Aldous, 1939). A specimen identified as *C. latrans thamnos* was taken in Lower Enchanted Township, Somerset County, Maine in 1961 (Carson, 1962). The subsequent occupation of nearly the entire state by the species has been documented by Richens and Hugie (1974) and Teer (1975). The Maine Department of Inland Fisheries and Wildlife (1976) has published an estimate of from 1,500 to 5,500 coyotes in the state.

Coyotes were first reported in Vermont in 1942 and have since been taken in all 14 counties of the state. The first record in New Hampshire was one shot near Holderness, Grafton County in 1944, and the first in Massachusetts was shot near Otis, Berkshire County in 1957. Three more were trapped in Massachusetts on the Prescott Peninsula of Quabbin Reservoir in 1958, and one was shot near Grafton, Worcester County in 1959. Four individuals, believed to be coyotes or hybrids between *C. latrans* and *C. familiaris* ("coydogs"), were taken in western Connecticut from 1957 to 1963 (Pringle, 1960, 1963). Silver and Silver (1969:Fig. 30) depicted occurrences throughout Vermont, New Hampshire, and Massachusetts.

Difficulties in Identifying Northeastern *Canis*

Although the recent presence of wild *Canis* in the northeast does appear to represent primarily an extension of the range of the species *C. latrans*, the exact identity of certain individuals and populations has been open to question. Perhaps because coyotes had never before occurred in the region, their initial appearance was something of a mystery and the cause of excitement among the public and local wildlife officials. Some persons believed that wolves were returning to areas in which they had been exterminated long before, while others considered the new canids to be wild dogs or coy-dogs. And it is likely that each of these three kinds of animals contributed in some part to the mystery. An actual specimen of *C. lupus* was killed in Fulton County, New York in 1968 (Paradiso and Schierbaum, 1969). Another wolf, of unsually large size and probably not a native animal, was killed north of Kemptville, Carleton County, Ontario, about 30 miles from the New York border, in 1962. According to Bump (1941) an undetermined number of gray wolves escaped from captivity in southern Franklin County, New York about 1930. Silver and Silver (1969) said that four gray wolves were imported to the vicinity of Croydon, Sullivan County, New Hampshire, and that after the last died in 1914 there were regular reports of wolflike animals in that area.

Completely feral dogs seem to be uncommon, but do occur on occasion, and are sometimes the cause, directly or indirectly, of reported wolves and coyotes. Carson (1962) referred to several cases of *C. familiaris* living and breeding in the wild in Maine. McKnight (1964:48) reported that a pack of wild dogs inhabited a den near Hopkinton, New Hampshire for six years prior to 1959. And Nesbitt (1975) made a five-year study of a feral pack on Crab Orchard National Wildlife Refuge, Illinois.

Aldous (1939) and Carson (1962) referred to the taking of numerous coy-dogs in Maine. Wetzel and Penner (1962) reported the collection of two specimens of coy-dogs in Litchfield County, Connecticut. Cook (1952) discussed the presence of such animals in New York. Wolfram (1964) cited a report that 20 percent of the coyotes in Ontario were actually coy-dogs. Paul (1970) stated that 20 percent of the carcasses of wild *Canis* found in Illinois were identified as coy-dogs. In tracing the spread of coyotelike animals in New York, Severinghaus (1974a, 1974b) noted that the first litters in a newly occupied area were almost always obvious coy-dogs.

Such records apply to wild-caught animals that were presumed to be hybrids between *C. latrans* and *C. familiaris* on the basis of morphological characters. Cases in which known hybrids were born in captivity under controlled conditions were discussed by Dice (1942), Young (1951:123), Kennelly and Roberts (1969), Gier (1968), and Mengel (1971). The latter three of these authors reported such hybrids to be fertile.

Mengel (1971) reviewed the subject of hybridization between *C. latrans* and *C. familiaris*, and presented information on his own experiments. He noted that whereas coyotes normally mate from late January to March, and usually give birth in the spring, coy-dogs have been observed to mate from October to December, and to give birth in the winter. This phase shift in the breeding cycle was held to be a barrier restricting the interbreeding of coy-dogs with *C. latrans*, and hence preventing the introgression of domestic dog genes into the wild coyote population. Furthermore, Mengel pointed out that since the offspring of coy-dogs would be born under harsh winter conditions, and since male coy-dogs do not demonstrate the same tendency to parental care as male coyotes, the hybrid pups would be unlikely to survive. These factors taken together seemingly would prevent the establishment of a population of canids of mixed coyote and dog ancestry. Gier (1968) and Kennelly and Roberts (1969) also reported the shift in breeding time among coy-dogs. Iljin (1941) found a parallel situation in captive hybrids of *C. lupus* and *C. familiaris*.

Gipson (1972), Gipson, Sealander, and Dunn (1974), and Gipson, Gipson, and Sealander (1975) concluded that in Arkansas introgression of dog genes into the coyote population could and did occur. This view was based in part on the fact that 38 of 284 skulls of wild canids recently collected in Arkansas, were shown by multivariate analysis to be morphologically intermediate to dog and coyote populations, and thus were designated coy-dogs. There was no evidence, however, that any of these individuals were other than first generation offspring. It also was found that some male coyotes in Arkansas were physiologically in breeding condition in late November and December. On the basis of examination of female reproductive tracts, however, the earliest reported actual mating was 17 February.

Freeman (1976:40-42) pointed out that in Oklahoma, at least, winter birth would not seriously affect the offspring of coy-dogs, since weather conditions were not so severe in that area. He also suggested that lack of care by the male parent would not necessarily result in loss of the litter, since female coyotes often successfully rear young after losing their mate.

Silver and Silver (1969) raised and studied a litter of five coyotelike canids dug from a den in Croydon, Sullivan County, New Hampshire, and 50 of their descendents. Two hybrid litters were born to one of the original females crossed with domestic dogs. The hybrids were fertile and produced two litters, but once again it was found that breeding occurred three to four months earlier than in wild *Canis*, and that males did not assist in rearing the young. Considering these factors, and also the disadvantages faced by pups born in midwinter, Silver and Silver thought it unlikely that coy-dogs could establish or merge with a wild population. On the basis of behavioral and physical studies of these various animals, it was concluded that the present population of New England *Canis* is of predominantly coyote ancestry, but that some dog and/or wolf genes had been introduced in the past. Such genes, however, had become well integrated so that the population was now stabilized and breeding true to type. Therefore, Silver and Silver thought that wild *Canis* in New England should not be considered as hybrids, but as a kind of coyote.

Lawrence and Bossert (1969) subjected 31 skulls of New England *Canis* to linear discriminant analysis. Of these specimens, 16 were offspring of the captive canids studied by Silver and Silver, and 15 were killed wild in New Hampshire, Vermont, and Massachusetts. In the analysis, most of the specimens fell between known samples of *C. lupus* and *C. latrans*, a few overlapped with *C. latrans*, and a few also approached the range of *C. familiaris*. The positions of these wild individuals were, however, substantially different from those of 21 known, captive raised F_1 and F_2 coyote-dog hybrids. Lawrence and Bossert, like Silver and Silver, concluded that the New England population was predominantly coyote, probably with some dog/wolf ancestry. A series of 32 skulls of *C. latrans thamnos* from Minnesota also demonstrated a shift away from typical *C. latrans* toward both *C. familiaris* and *C. lupus*, but not to the extent found in New England *Canis*. Hence the New England population was considered to represent the development of a trend that had begun at an earlier time through the introduction of wolf or dog genes into the coyote population. The designations "*Canis latrans* var." and "eastern coyote" were used for the wild population of New England *Canis*.

Chambers, *et al.* (1974) used the same technique as Lawrence and Bossert (1969) on nearly 150 recently collected specimens from New York, and came to the same conclusion regarding the systematic position of the population. Still another statistical analysis of skulls was done on Maine material by Hilton (1977) who found that, except for five dogs, all specimens received from 1968 to 1975 could be identified as eastern coyotes.

As previously mentioned, Mengel (1971) thought that the introduction of dog genes into a population of wild *Canis* would be unlikely. He therefore suggested that the present population of New England *Canis* was the result of the introgression of genes only from the wolf (*C. lupus lycaon*) into the wild coyote population. This view was supported in part by a discriminant function analysis in which skulls of 55 coyotes taken in northeastern Kansas, 12 known captive raised coyote-dog hybrids, and 13 presumptive wild coy-dogs were compared to the same samples of *Canis* used by Lawrence and Bossert (1969). Since domestic dogs are abundant in northeastern Kansas, and since coy-dogs are regularly reported in the area, Mengel argued that the local coyote population should demonstrate the same variability as that of New England, if introgression from *C. familiaris* were possible. Because in the analysis the 55 coyotes fell close to the original coyote sample, and because the positions of the coyote-dog hybrids substantially differed from those of the Kansas coyotes, Mengel concluded that introgression from *C. familiaris* to *C. latrans* was not occurring. *Canis lupus*, long extinct in Kansas, but still present in parts of the northeast, was thought to be the only source of the genes that had modified the northeastern coyote population.

Known cases of hybridization between *C. lupus* and *C. latrans* are rarer than those between other pairs of species of *Canis* (Gray, 1972). Indeed, until recently the only suggestion that such a cross was possible was the statement by Young (1951:124) that two specimens of supposed coyote-wolf hybrids, born in captivity, were in the Royal Ontario Museum. Kolenosky (1971) reported that in May 1969 and May 1970, litters of five healthy pups were born to a female *C. lupus lycaon*, captured in Algonquin Provincial Park, and a male *C. latrans*, taken in York County, Ontario. Kolenosky and Standfield (1975) added that a subsequent mating of F_1 siblings produced a litter of four pups. They also observed that the members of the F_1 generation were similar in appearance to many specimens of a small kind of wolf, designated the "Tweed type," that had been collected along the southern limits of the range of their "Algonquin type" of *C. lupus lycaon* in southeastern Ontario. They stated that evidence was mounting that the "Tweed type" had originated from hybridization between *C. latrans* and *C. lupus lycaon* in the wild.

Examination of Specimens

With the above background in mind, I examined all skulls of wild *Canis* from the northeastern United States and southeastern

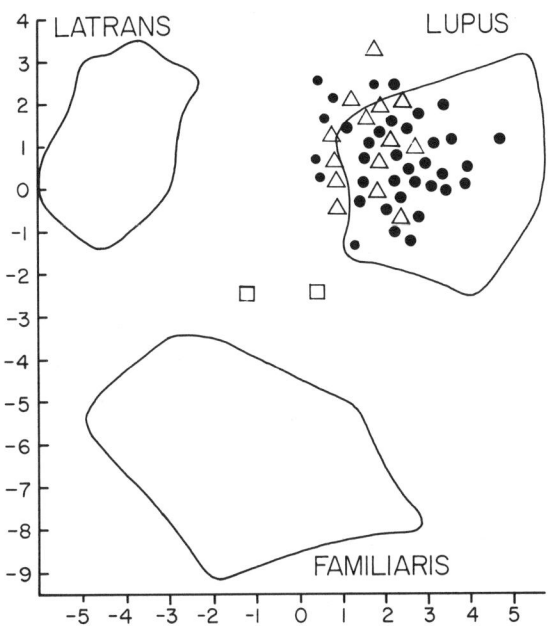

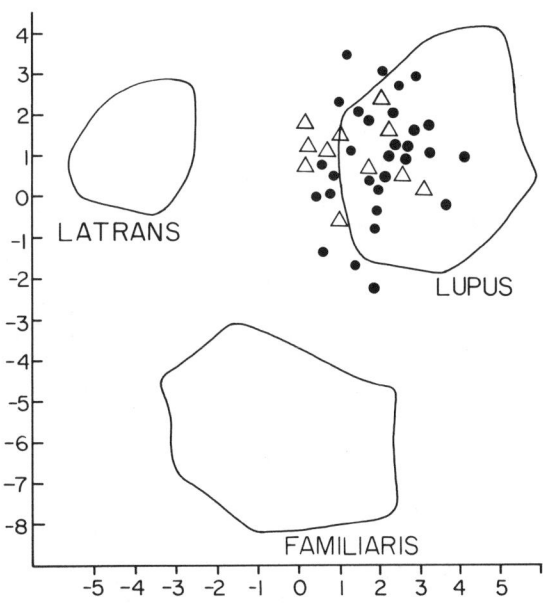

Fig. 7.—Multivariate comparison of individual specimens of *C. lupus lycaon* to the ranges of variation of the total series shown in Fig. 5. Black dots, statistical positions of *lycaon* from Michigan, Minnesota, Wisconsin, and western Ontario; triangles, positions of *lycaon* from southeastern Ontario and southern Quebec; squares, positions of two specimens thought to be hybrids between *C. lupus lycaon* and *C. familiaris*. Males are above, females below.

Canada that were available to me. The previously established total series of northern and western wolves and coyotes, and the series of 50 domestic dogs, were used as standard groups against which the northeastern material was tested (see pp. 11-12).

Each skull previously identified by others as *C. lupus lycaon,* except the six to be discussed later, were compared individually to the three standard groups by multivariate analysis. Most of these skulls fall within the range of variation of *C. lupus,* but, as could be expected considering their relatively small over-all size and narrow proportions, they demonstrate a shift toward *C. latrans* (Fig. 7). Two males from the upper peninsula of Michigan (see appendix A, part 6) have a statistical position intermediate to those of *C. lupus* and *C. familiaris,* and also appear by eye to be of mixed blood. These two skulls, therefore, are considered to represent hybrids, and henceforth are not used in the analyses.

Of the remaining specimens represented in figure 7, a group of 72 were collected in the upper peninsula of Michigan, northern Wisconsin, northern Minnesota, and that part of Ontario to the north and west of Standfield's (1970) line separating the "Algonquin type" and "Ontario type" of *C. lupus lycaon;* and a group of 31 were obtained in southeastern Ontario and southern Quebec (see appendix A, part 7, and Fig. 2). The multivariate distribution of these two groups shows overlap and does not suggest any sharp distinction. Wolves from the more westerly region are larger, and have relatively broader rostra and frontal shields, and smaller second upper molars (see appendix B, part 3). In these and other characters the western *lycaon* do appear to fall between the eastern *lycaon* and *C. lupus nubilus* from the Great Plains. Thus Mech and Frenzel (1971) could have a case in suggesting the survival of *nubilus* in Minnesota. Throughout the entire region in question, however, we seem to be dealing

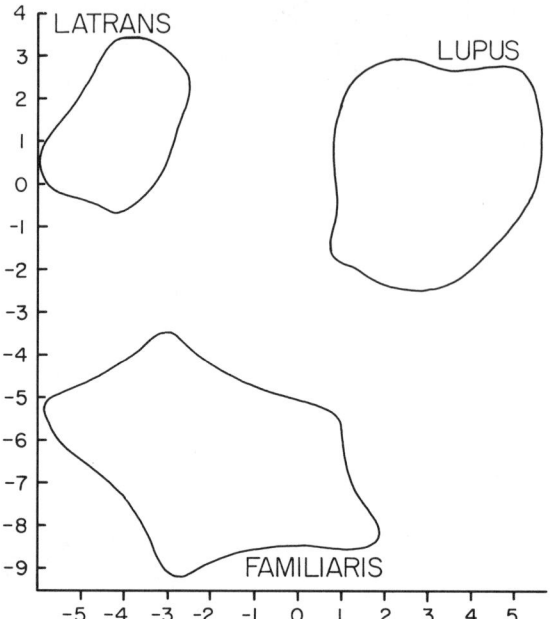

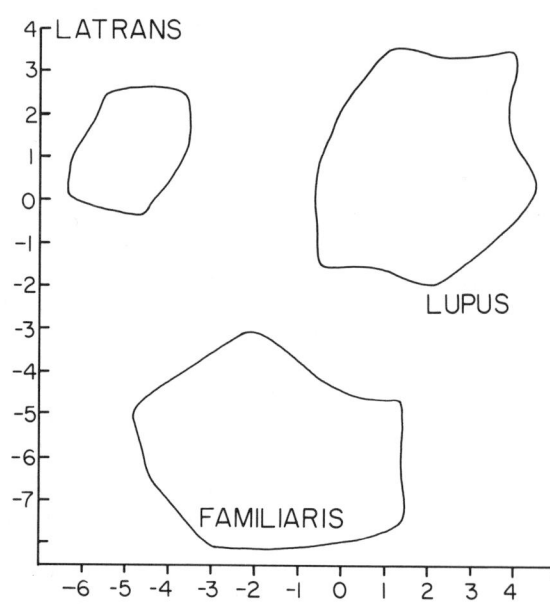

Fig. 8.—Multivariate comparison of all specimens of *C. lupus* (including *lycaon*), specimens of *C. latrans* from northern and western North America, and *C. familiaris*. Only the margins of the range of variation of each species are shown. Males are above, females below.

only with minor and gradual variation, when considering the species *C. lupus* as a whole.

These findings are not intended to contradict the conclusion of Kolenosky and Standfield (1975) that there is a significant statistical difference between specimens of *C. lupus lycaon* from various parts of Ontario. Those authors dealt primarily with a question of intraspecific variation. My studies tend to show only that material referred to *lycaon* is not more noticeably distinctive from *C. lupus* in general than is material referred to most other named subspecies of the gray wolf.

Satisfied that the 103 skulls of *lycaon* discussed above were representative of the species *C. lupus*, I incorporated them into my gray wolf samples. Graphical results of multivariate analyses comparing these total samples with the standard samples of domestic dogs and coyotes are shown in figure 8.

All available skulls, previously identified in collections as coyotes (with one exception to be discussed later), and that had been taken in the range delineated by Jackson (1951:266-267) for *C. latrans thamnos* (plus extreme southern Quebec), were individually tested against the three standard groups of *C. lupus* (including *lycaon*), *C. latrans*, and *C. familiaris*. The specimens included 80(22) males and 50(13) females from Manitoba, North Dakota, Minnesota, Iowa, Illinois, Indiana, Michigan, Wisconsin, Ontario, and Quebec (see appendix A, part 8). Relative multivariate positions are depicted in figure 9. Nearly all specimens fall within or near the total range of variation of *C. latrans*, but are concentrated closer to *C. lupus* than is the standard coyote sample. One skull, obtained in March Township, Ontario, appears to represent a coyote-dog hybrid. The overwhelming statistical affinity of *thamnos* to the standard sample of *C. latrans* suggests that the coyote has established itself in essentially unmodified form in much of the northeast.

In addition to the above, I was able to examine a number of specimens from the extreme northeastern United States where the wild canid population was designated

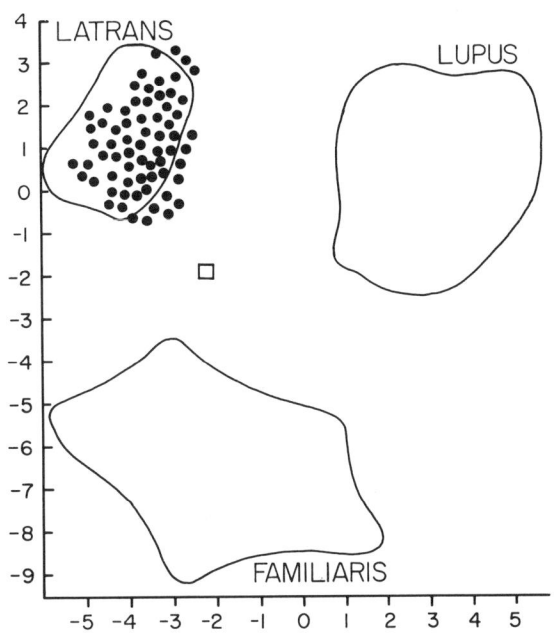

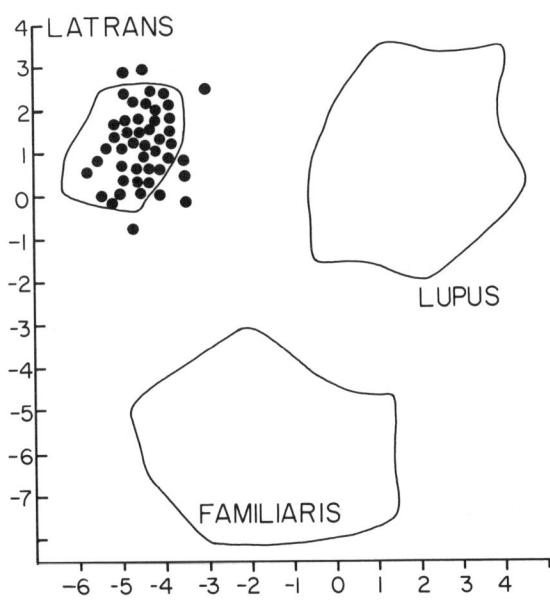

Fig. 9.—Multivariate comparison of individual specimens of *C. latrans thamnos* to the ranges of variation of the total series of *C. lupus*, *C. latrans*, and *C. familiaris* shown in Fig. 8. Black dots, statistical positions of *thamnos*; square, position of probable hybrid between *C. latrans thamnos* and *C. familiaris*. Males are above, females below.

"*C. latrans* var." by Lawrence and Bossert (1969). Wild-caught animals were represented by skulls of 25(5) males and 20(2) females (see appendix A, part 9). Multivariate positions, relative to the three standard groups, are depicted in figure 10.

Figure 10 also shows the statistical positions of nine known coyote-dog hybrids (some of those discussed by Mengel, 1971; all in KU), and the positions of 15 wild-caught individuals (all of those in USNM and KU that could be utilized) previously identified by others as suspected coy-dogs.

The specimens of wild northeastern *Canis* are statistically proximal to the standard sample of *C. latrans,* but demonstrate a pronounced shift toward *C. lupus.* By eye, most of these skulls appear coyotelike, though on the average they are larger than *C. latrans* and have relatively broader rostra and frontal shields. Both visually and statistically, five individuals stand out from the others and have multivariate positions approaching the limits of the sample of *C. familiaris.* My data thus support the view of Lawrence and Bossert (1969) that both domestic dog and gray wolf genes have influenced *Canis* in the northeast. Nonetheless, the amount of graphical separation between the main clusters of northeastern canids and the few that are scattered in the direction of *C. familiaris,* suggests to me that two separate phenomena are involved. First, a limited amount of hybridization between wild *Canis* and domestic dogs has occurred, but there has been no large-scale shift in characters, among the wild population as a whole, in the direction of *C. familiaris.* Secondly, the predominantly coyotelike population of northeastern *Canis* evidently has sustained the introgression of genes from the gray wolf. There are not yet enough data to determine exactly when and where this process began, but apparently wolf genes are now incorporated in the northeastern coyote.

The relationships among the populations of *Canis* in the northeast and in other regions

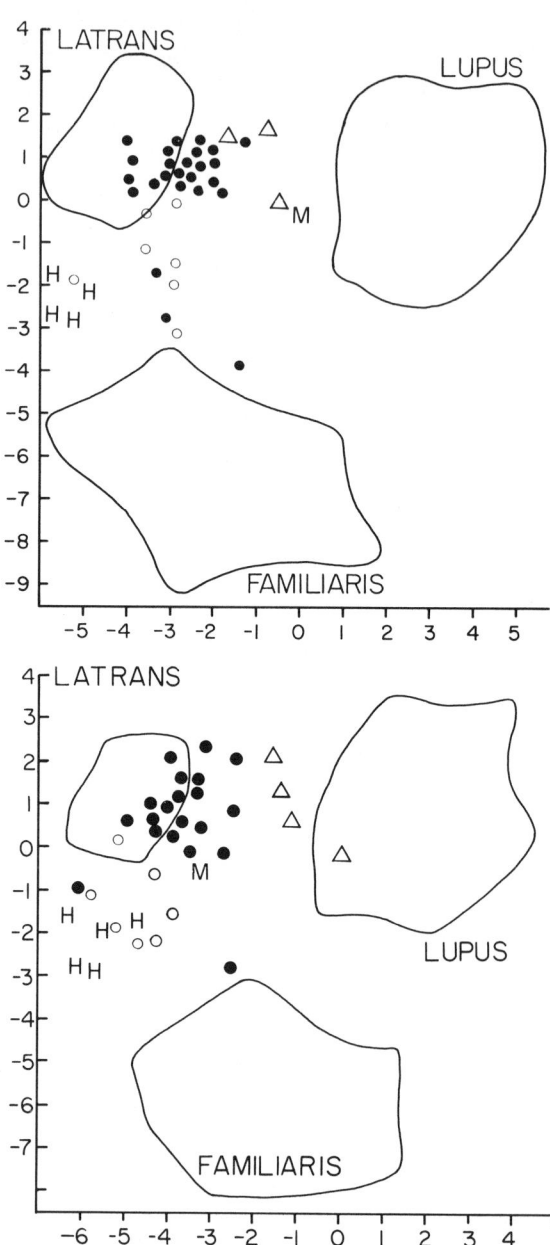

Fig. 10.—Multivariate comparison of various individual specimens to the ranges of variation of the total series of *C. lupus, C. latrans,* and *C. familiaris* shown in Fig. 8. Black dots, statistical positions of specimens of wild-caught individuals from New England, New York, and Pennsylvania; open circles, wild-caught animals previously reported as coy-dogs; H, captive born coy-dogs; triangles, specimens from southeastern Canada thought to be hybrids between *C. lupus* and *C. latrans;* M, specimens in ROM thought to be from captive wolf-coyote hybrids. Males are above, females below.

TABLE 1
Statistical distance, D^2, between populations of *Canis* in the northeast and elsewhere.

	C. familiaris	C. lupus	C. lupus lycaon	C. latrans	C. latrans thamnos
A. Males					
C. lupus	68.2				
C. lupus lycaon	61.7	4.1			
C. latrans	55.8	68.1	48.3		
C. latrans thamnos	53.3	63.2	42.2	2.0	
Northeast U.S. Canis	48.4	49.2	30.2	5.4	3.2
B. Females					
C. lupus	51.6				
C. lupus lycaon	46.0	4.6			
C. latrans	56.6	56.8	39.3		
C. latrans thamnos	55.8	53.6	34.9	2.2	
Northeast U.S. Canis	47.4	41.3	25.1	3.5	2.1

hitherto discussed, are summarized in table 1. Data from those specimens considered to represent hybrids between wild *Canis* and *C. familiaris* were not used in the calculation of the statistics shown. Northern and western wolves demonstrate close relationship with the subspecies *C. lupus lycaon* of the northeast; and there is affinity between northern and western coyotes, *C. latrans thamnos,* and the populations of *Canis* in the extreme northeastern United States. This latter group and *C. lupus lycaon* are separated by less than half the statistical distance found between the main groups of northern and western wolves and coyotes.

I above postponed discussion of six specimens previously identified in collections as *C. lupus lycaon,* and one identified as *C. latrans thamnos* (see appendix A, part 6). Each of these skulls appeared to be of unusual size or to have other characters suggesting hybridization between *C. lupus* and *C. latrans.* Multivariate positions are shown in figure 10.

In addition to the above, I examined two skulls (ROM 31-9-15-1 and 31-9-15-2) reported to represent the hybrid offspring of a captive wolf and coyote (Young, 1951:124; Mengel, 1971:333). The labels of these specimens read in part: "From Ontario stock." Multivariate positions of each, relative to the three standard series, are plotted in figure 10 (since ROM 31-9-15-2 lacked both P1, it would not ordinarily have been included in a multivariate analysis, but because of its special interest I used an estimate of alveolar length of upper toothrow).

If considered together, the positions of these nine specimens bridge the gap between the statistical limits of the standard sample of *C. lupus* and the positions of the individual specimens from the northeastern United States. It is thus reasonable to suppose that hybridization between wolf and coyote in southeastern Canada has, and probably still is permitting the flow of genes from one species to the other. Evidently wolf genes have spread through much of the coyote population, and resulting wolflike characters have been phenotypically expressed in *Canis* of the extreme northeastern United States. This introgression may have assisted the coyote population of the region to adapt to and flourish in an environment far from the original prairie habitat of *C. latrans.*

Systematic Problems in the Southeast

Background

When Europeans first entered the southeastern part of North America, they found animals that appeared to be closely related to the wolves of the Old World. For example, Catesby (1771:xxvi), writing of Florida and the Carolinas, noted: "The Wolves in America are like those of Europe, in shape and colour, but are somewhat smaller."

Bartram (1791:199), the first author to apply a binomial to *Canis* in the southeastern region, referred to the black wolves that he saw in Florida as "lupus niger." Harlan (1825:82) used the name *Canis lycaon* for these same animals, but also for black wolves inhabiting mountainous areas of North America and Europe. The designation *Canis lycaon* originally had been given to a black wolf from the vicinity of Quebec by von Schreber in 1775 (Goldman, 1944:437–440), and was restricted by Miller (1912b) to the wolves of eastern Canada and the northeastern United States. Richardson (1829:70) termed the black wolves that he saw on the banks of the Mackenzie and Saskatchewan rivers of western Canada as "*Canis lupus occidentalis* var. E. *Lupus ater.*" He said, however, that the same animals occurred throughout North America, and he included Bartram's Florida wolves under this name. It thus seems that early naturalists named kinds of wolves largely on the basis of color, and considered the names to apply wherever the particular colors were found. Since it was eventually demonstrated that coloration of wolves in nearly all parts of North America is highly variable and of minimal taxonomic value (Young, 1944:59–66; Goldman, 1944:401), these early writings are useless to an understanding of the problems presently under review.

Apparently Audubon and Bachman (1851: 126, 240) were the first authors to set definite bounds to the ranges of named kinds of North American wolves, and to suggest that in the southern United States there existed wolves structurally different from those in other regions. They kept Richardson's designation in their description of *Canis lupus* var. *Ater*, the "Black American Wolf," but recorded this kind only from Florida, South Carolina, North Carolina, Kentucky, southern Indiana, southern Missouri, Louisiana, and northern Texas. They also discussed *Canis lupus*, var. *Rufus,* the "Red Texan Wolf," which they said ranged from northern Arkansas, through Texas, and into Mexico. In their description of this animal Audubon and Bachman mentioned the long legs, pointed nose, and slender proportions. They noted that although the reddish shade predominated in Texas, other colors also were represented there, and that the wolves of different colors freely interbred. They stated that except for *Canis latrans,* the coyote, all of the wolves that they described were only varieties of one species.

Audubon and Bachman's delineation of *Canis* in the southern United States was generally accepted by biologists. But toward the end of the nineteenth century finer taxonomic splitting became more fashionable, and Bangs (1898) designated the Florida wolf as a full species, *Canis ater.* Bailey (1905) questionably referred to the large, dark wolf of east Texas as *C. ater,* and expressed hope that specimens could be obtained to confirm its status. He also recognized Audubon and Bachman's red wolf as a full species with the name *Canis rufus,* and assigned it a range in southern and central Texas. Bailey distinguished the larger gray wolf (*C. griseus*= *C. lupus*) of western Texas from *C. rufus,* and referred to the latter as "a large coyote or small wolf."

Miller (1912a) explained that the name *ater* was technically unavailable, and he designated the Florida wolf as *C. floridanus.* This name then became generally accepted for wolves in the forested areas of the southeastern United States, while *C. rufus* continued to be recognized in central and southern Texas. At this time, however, there had not yet been an attempt to associate *C. rufus* and *C. floridanus* into one group that demonstrated characters different from those of all other North American wolves.

Goldman (1937) combined the wolves of the south-central and southeastern United States into a single species, *C. rufus*, that he considered distinct from all other North American wolves. He combined the latter into the species *C. lupus*. Goldman said that *C. rufus* "exhibits a departure from the true wolves, and in cranial and dental characters approaches the coyotes." He listed the names *C. rufus rufus* for the Texas subspecies, and *C. r. floridanus* for the eastern race, and he also described *C. r. gregoryi*, a new subspecies in the lower Mississippi Valley. Shortly thereafter, Harper (1942) pointed out that Bartram's (1791) term for the Florida wolf was actually the earliest name to have been applied in the southeast, and the specific designation of the wolves in the region then technically became *C. niger*. But the International Commission on Zoological Nomenclature (1957, opinion 447) rejected Bartram's technical terms, and Hall (1965) listed *C. rufus* as the proper name. Therefore, the trinomials applied by Goldman (1937) are presently considered valid and are used in subsequent discussion.

Goldman (1944:481) observed that the two eastern subspecies of *C. rufus* exhibited a remarkable approach in size and general proportions to the eastern gray wolf, *C. lupus lycaon*, but that there existed several specific cranial differences. On the other hand, he said that the subspecies *C. r. rufus* in central Texas and Oklahoma was so small and in general characters agreed so closely with *C. latrans*, that some specimens were difficult to distinguish. He suggested the possibility of hybridization between *C. rufus* and *C. latrans* in some localities in Texas.

Goldman (1944) used the vernacular "red wolf" for the species *C. rufus*, presumably on the basis of Audubon and Bachman's description of the "Red Texan Wolf." This term, however, is not found in any of the early literature discussing wolves in states east of Texas. Kellogg (1915:41) said that trappers in Cherokee County, southeastern Kansas, did refer to "Red Wolves" in the area. But otherwise this popular term seems to have been restricted to parts of Texas until Goldman introduced its use throughout the range of *C. rufus*. This appellation may be unfortunate, because although the rufous element in the fur sometimes stands out, the "red barn-roof paint" color mentioned by Young (1946:36) seldom shows up. Early records indicate that a dark-colored or entirely black phase was locally common in the eastern forests. But most available specimens of "red wolves" actually exhibit a typical wild canid color pattern, consisting of an agouti gray or brown, interspersed with black hairs, especially on the back, and with the muzzle, ears, and outer surfaces of the limbs tending toward a tawny color.

Lawrence and Bossert (1967) thought that separation of *C. rufus* as a distinct species rested too heavily on the small red wolves of central Texas where hybridization with the coyote may have been a factor. They said that if the study of wolves in the south had been based on adequate series of specimens from Florida, separation of *C. rufus* from *C. lupus* would have been highly unlikely. In a multivariate analysis they compared all available skulls of wolves collected before 1920 in Louisiana, Alabama, and Florida, including one Florida specimen assigned by Goldman to *C. lupus lycaon* (according to Lawrence and Bossert, 1975:81, a total of 12 specimens were in this sample), with series of 20 adult skulls each of *C. familiaris*, *C. lupus* (various North American subspecies), *C. lupus lycaon*, and *C. latrans*. They found *C. lupus lycaon* and *C. rufus* (the early southeastern material) both to overlap with *C. lupus*, but to be distinct from each other. All three of these populations formed a cluster distinct from *C. latrans*, with *C. rufus* being the farthest removed. Lawrence and Bossert interpreted this analysis as demonstrating that early populations described as *C. rufus,* "east of the range of *Canis latrans,* are a local form of *Canis lupus,* not a distinct species of wolf."

Lawrence and Bossert (1975) repeated the above interpretation. They again observed that their 12 early specimens from Louisiana and eastward, which they now designated *"floridanus,"* and their series of *C. lupus lycaon,* were both less coyotelike than their general sample of *C. lupus.* Lawrence and Bossert (1975) also evaluated a series of 30 specimens (designated as *"gregoryi"*) taken in the 1920's in Arkansas, south of the Arkansas River. This group was found to be nearer in D^2 distance to the general sample of *C. lupus* and to *C. lupus lycaon,* than to *"floridanus."*

The suggestions by Lawrence and Bossert (1967, 1975)—(1) that pre-1920 eastern *C. rufus* is not more than subspecifically distinct from *C. lupus,* (2) that *C. rufus* and *C. lupus lycaon* are less coyotelike than *C. lupus* in general, and (3) that Arkansas red wolves are closer to *C. lupus* than to Louisiana red wolves—are not in agreement with my own findings. These suggestions, at least in part, also seem not to correspond well with results obtained by some other workers.

Paradiso (1968) argued that the relatively small samples used by Lawrence and Bossert (1967) did not adequately represent the variability shown by *Canis,* and thus that conspecificity of the red and gray wolves had not been demonstrated. He also pointed out several cranial and dental characters in which *C. rufus* resembled *C. latrans* more than it did *C. lupus.* Paradiso and Nowak (1972a) compared data on 213 skulls of *C. rufus,* 214 of *C. lupus,* and 336 of *C. latrans,* and concluded that the red wolf was a distinct species. In size and proportion *C. rufus* fell between *C. latrans* and *C. lupus,* but was nearer to the latter. Development of certain dental characters in *C. rufus* suggested affinity to the coyote. Atkins and Dillon (1971), on the basis of a morphological study of the cerebellum, considered the red and gray wolves to be in the same group, distinct from other *Canis,* but that *C. rufus* was a valid species. Shaw (1975) evaluated data on the behavior, ecology, vocalizations, allelic frequency, and morphology of a living population of *C. rufus gregoryi* in southeastern Texas, and concluded that the population represented a species distinct from both *C. lupus* and *C. latrans.*

Three recent studies, aimed primarily at identifying newly collected specimens of wild *Canis* in the south-central states, and each employing a different method of multivariate analysis, also have tended to uphold the specific status of *C. rufus.* Gipson, Sealander, and Dunn (1974), who used a single measurement and five separate ratios as variables, compared as groups the skulls of 40 *C. latrans,* 34 *C. latrans* x *C. familiaris,* 31 *C. familiaris,* 37 *C. lupus lycaon,* 40 *C. lupus* from the Great Plains region, and 40 *C. rufus* taken in Arkansas prior to 1925. They found the red wolf sample to be statistically intermediate to the coyote and gray wolf samples, and reported their results to support recognition of the red wolf as a distinct species. Freeman (1976), who used 15 raw measurements as variables, compared as groups the skulls of 40 *C. latrans,* 12 *C. latrans* x *C. familiaris,* 24 *C. familiaris,* 43 *C. lupus* from the Great Plains region, and 44 *C. rufus* taken in Arkansas prior to 1925. The graphical results of his analysis indicate that the red wolf, again, is statistically between the coyote and gray wolf. Unlike Gipson, Sealander, and Dunn (1974), however, who found some overlap between samples of *C. lupus* and *C. rufus,* Freeman obtained complete statistical separation of the two species. Elder and Hayden (1977), whose variables were proportions obtained by dividing 14 measurements by greatest length of skull (in accordance with Lawrence and Bossert, 1967), compared as groups the skulls of 29 *C. latrans,* 30 *C. familiaris,* and 18 *C. lupus* from widely scattered regions. Then,

27 skulls of *C. rufus* taken in Missouri in the 1920's and early 1930's were compared as individuals to these three groups. None of the red wolves fell within the limits of the groups, but they were distributed statistically between the coyote and gray wolf samples.

Mech (1970:25, 285, 351) suggested that the differing views on the status of *C. rufus* could be reconciled by recognition of the red wolf as a hybrid population that could be properly known as *C. lupus* x *C. latrans*. Regardless of the taxonomic disputes, it is now generally agreed that the original populations of *Canis* inhabiting the southeastern quarter of the United States have all but disappeared, and have been replaced in large part by another kind of *Canis* (McCarley, 1962; Paradiso and Nowak, 1972a; Pimlott and Joslin, 1968).

Examination of Earliest Available Eastern Material

In my own study, I wished to consider all available specimens from south of Lakes Michigan and Erie, and east of the Mississippi River. Goldman (1944) examined only four skulls (including one damaged and one immature) and one mandibular ramus, assigned to *C. lupus lycaon*, from this region. He also had only two skulls of *C. rufus gregoryi* (both subadults), from Indiana and Illinois, and two of *C. r. floridanus* (one damaged), from Alabama and Florida. Some other workers have considered this small collection to be insufficient for an assessment of the relationships among the original populations in the region, particularly for settling the question of whether *C. rufus* and *C. lupus* intergraded in the Ohio Valley and along the central Atlantic coast. Unfortunately, there are scarcely any additional skulls available, and there doubtless never will be, because the native wolves of the region were exterminated long ago. Excavation of fossil and aboriginal sites offers some hope, but most material from such sources that I have examined is frag-

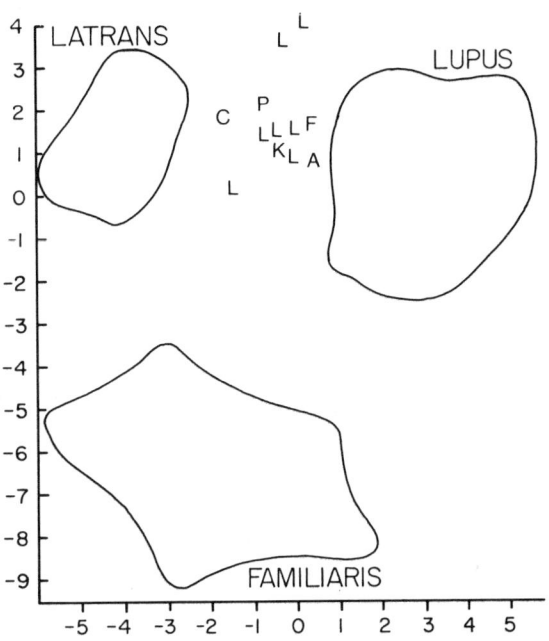

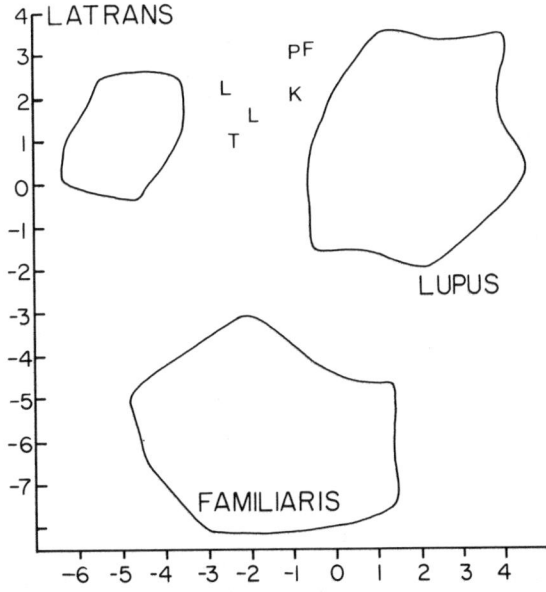

Fig. 11.—Multivariate comparison of individual specimens of *C. rufus* to the ranges of variation of the total series of *C. lupus*, *C. latrans*, and *C. familiaris* shown in Fig. 8. L, pre-1920 specimens from Louisiana; A, specimen taken in 1917 in Alabama; C, specimen found in Alabama cave deposit; F, Florida specimen; K, specimen from Garvin County, Oklahoma; P, Pennsylvania specimen; T, southeastern Texas specimen. Males are above, females below. The Florida, Oklahoma, and Pennsylvania specimens are compared both with male and female series.

mentary and of limited value. This material is discussed later in the paper.

Lawrence and Bossert (1967) limited their sample of *C. rufus* to specimens collected before 1920, believing that such material was the least likely to show influence from *C. latrans*. As an initial step, this paper also considers material taken prior to 1920. But from this period, and from east of the Mississippi River, there are available only four skulls that are adult and unbroken, and hence suitable for inclusion in multivariate analysis. Two of these specimens are unknown as to sex, but since the smallest of the four is a known male, I compared each of them individually with the previously compiled samples of 50 *C. familiaris*, 166 male *C. latrans*, and 294 male *C. lupus* (including 61 *C. lupus lycaon*). Their relative multivariate positions are plotted in figure 11 (the two skulls of unknown sex are also compared to appropriate female samples). The positions of the four are beyond the range of variation of any of the known samples, but are distributed between *C. lupus* and *C. latrans*. Measurements of these skulls are listed in appendix B (part 4), and other details are given in table 2.

Although these four specimens are indicative of the former presence in the eastern United States of a kind of wolf different from any subspecies of *C. lupus*, their value is limited by missing data. A better collection of pre-1920 material was taken in Louisiana, and was identified by Goldman as *C. rufus gregoryi*. The Louisiana material suitable for inclusion in my analysis consists of 7(2) males and 2(1) females (parentheses contain numbers of specimens in the series for which sex had not been recorded, but which were judged to belong to the particular sex indicated; see p. 5). A single skull of a female *C. rufus gregoryi*, taken in extreme southeastern Texas in 1906, also is available.

As depicted in figure 11, the nine Louisiana specimens, and the one from southeastern Texas, have multivariate positions intermediate to the graphical limits of *C. lupus* and *C. latrans*. Seven of these skulls have individual D^2 distances farther from *C. lupus*

TABLE 2

Data on early specimens of *C. rufus*.

Collection and number	Sex	Locality	Date	D^2 from *C. lupus*	D^2 from *C. latrans*
ANSP 2261	male?	Pennsylvania	pre-1859	27.2	23.0
MCZ 11179	male?	Florida*	1854?	24.5	37.8
USNM 223936	male	12 mi. S Cherokee, Colbert County, Alabama	1917	14.5	30.6
USNM 348063	male	Fern Cave, Jackson County, Alabama**	----	43.8	26.4
USNM 132229	male	Mer Rouge, Morehouse Parish, Louisiana	1904	25.6	24.9
USNM 136834	male	23 mi. SW Tallulah, Madison Parish, Louisiana	1905	18.1	28.1
USNM 137125	male	20 mi. SW Vidalia, Concordia Parish, Louisiana	1905	37.9	27.2
MCZ 9114	male	Mer Rouge, Morehouse Parish, Louisiana	1898	23.2	22.8
USNM 136731	male	18 mi. SW Tallulah, Madison Parish, Louisiana	1905	46.1	52.3
USNM 133687	male?	15 mi. NW Tallulah, Madison Parish, Louisiana	1904	38.4	38.1
USNM 133688	male?	10 mi. SW Floyd, West Carroll Parish, Louisiana	1904	29.7	37.5
USNM 136105	female	15 mi. SW Tallulah, Madison Parish, Louisiana	1904	22.4	19.4
USNM 234227	female?	12 mi. N Avery Island, Iberia Parish, Louisiana	1919	31.7	19.1
USNM 147701	female	Kountze, Hardin County, Texas	1906	38.0	25.2

* Apparently with regard to this specimen, Barbour (1944:142) wrote: "we have in the Agassiz Museum in Cambridge a skull of a fine old adult animal which was said to have been killed in 1854 in the region where the city of Miami now stands."

** This specimen was found in a cave; its state of preservation suggested that it lived in Recent times (Paradiso and Nowak, 1973).

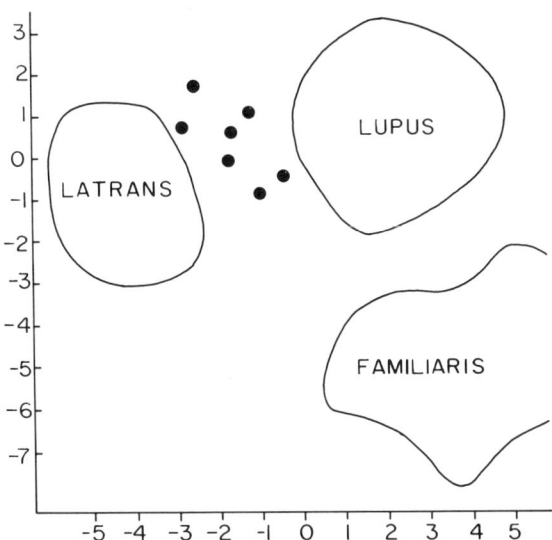

Fig. 12.—Multivariate positions (black dots) of individual pre-1920 males of *C. rufus* from Louisiana, relative to series of males of *C. lupus, C. latrans,* and *C. familiaris*. The variables used in this particular analysis were derived by dividing the measurements by greatest length of skull.

than from *C. latrans* (table 2). A separate analysis of males only, in which the employed variables were fractions derived by dividing each of the other measurements by greatest length (in accordance with Lawrence and Bossert, 1967), produced much the same picture (Fig. 12).

Measurements of the pre-1920 Louisiana material are listed in appendix B (part 4), and other details are given in table 2. In figure 13, specimens of males are compared with the series of *C. lupus* and *C. latrans* in a ratio diagram. The Louisiana skulls approach those of *C. lupus* in over-all length, but are comparatively small in most other dimensions, a notable exception being their large M2. In general the proportions of *C. rufus gregoryi* seem intermediate to those of *C. lupus* and *C. latrans,* but the frontal shield in the Louisiana material is even narrower, relatively, than in the coyote, and the postorbital constriction is narrower, relatively, than in the gray wolf. The results of these analyses indicate that *C. rufus* morphologically resembles *C. lupus lycaon* more than it does any other subspecies of gray wolf.

We are still left with what Lawrence and Bossert (1967) called "the biologically difficult problem of reconciling the existence of two similarly-sized forms of wolf in one continuous habitat." Actually, because of the limited number of specimens, there may never be indisputable proof that the red and gray wolves did not undergo intergradation in the eastern forests. The existence, however, of 14 complete skulls collected before 1920 in this region, that do not overlap in

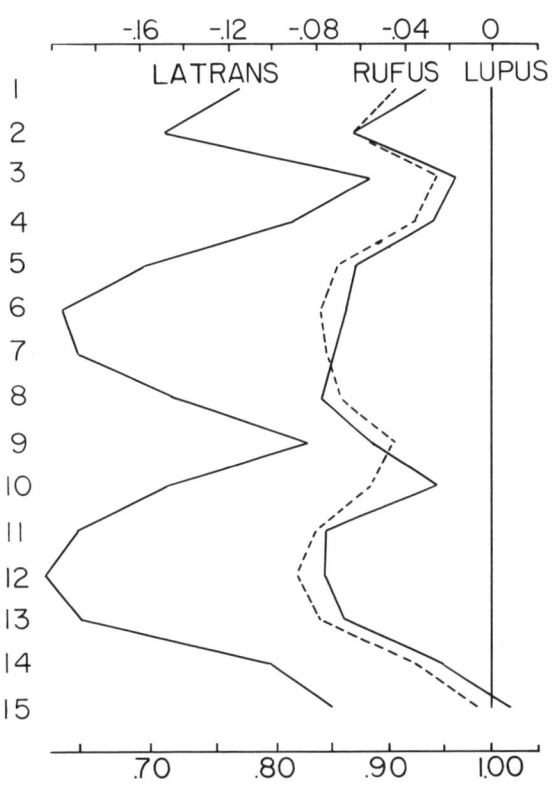

Fig. 13.—Ratio diagram comparing means of total series of *C. lupus,* series of *C. latrans* from northern and western North America, and two series of *C. rufus* (males only are shown for all series). The solid line under *rufus* represents the mean values of pre-1920 Louisiana specimens; the dashed line represents the mean values of specimens taken from 1919 to 1929 in the south-central states. The vertically arranged numbers correspond to the measurements so numbered in appendix B. A log difference scale is provided above, and a ratio scale below the diagram.

statistical position with series of 482 specimens of *C. lupus,* including 103 *lycaon,* suggests to me that the population represented by these skulls is more than subspecifically different from the gray wolf. Furthermore, the slender proportions of the red wolf indicate that its prey averaged smaller than that of the gray wolf, and that its ecological niche may have approached that of the coyote which did not exist in the eastern forests when white settlers first arrived. Possibly, when *C. lupus* entered the northeastern forests it underwent a degree of parallel evolution with *C. rufus,* just to the south. The subspecies *C. lupus lycaon* may even have been in the process of replacing *C. rufus,* when the white man interfered. And it is reasonable to think, considering the ease with which interbreeding occurs in *Canis,* that hybridization between *C. lupus lycaon* and *C. rufus* did occur, and that a zone of introgression may have developed that tended to modify one or both populations. The advent of the white man would probably have stimulated the spread of such hybridization, just as it seems to have encouraged interbreeding between *C. rufus* and *C. latrans* in Texas (Paradiso and Nowak, 1972a). Be that as it may, available early specimens from the southeastern United States can all be separated from known series of *C. lupus,* and seem to represent a different species, *C. rufus.*

The exact distribution of *C. lupus* and *C. rufus,* and the extent to which their ranges overlapped in eastern North America, will probably never be known. But if one Florida specimen, called *C. lupus lycaon* by Goldman (see table 2), is in fact *C. rufus* as I have tried to show, then there is no longer any confirmation of the original presence of the gray wolf in Florida. And if correct, identification of a Pennsylvania specimen as *C. rufus* rather than *lycaon* means that the range of the red wolf once extended farther to the northeast than was formerly thought.

As explained by Paradiso and Nowak (1972a:7-8), Goldman was also incorrect in his reasons for stating that the range of the red wolf in the Mississippi Valley once extended as far north as Warsaw, Hancock County, Illinois; and Wabash County, Indiana. The specimen supposedly from the former locality had been in possession of an animal dealer, and there is no evidence that it represents a native wolf of the area. The other specimen was actually collected in the Wabash River area of southwestern Indiana, not in Wabash County father north. Both skulls appear to be referable to *C. rufus,* but they are subadults and hence not suitable for use in multivariate comparisons made on the basis of adults.

Examination of Material Collected from 1919 to 1929 in the South-central United States

In the course of Federal predator control work in the south-central United States, from 1919 to 1929, a large number of specimens of *Canis* were taken. Most of these were referred by Goldman (1944) to *C. rufus gregoryi,* the "Mississippi Valley Red Wolf." Lawrence and Bossert (1967) did not use this material in their attempt to define the red wolf, because of the possibility that some of the specimens represented hybridization between red wolf and coyote. Paradiso and Nowak (1972a), however, noted that although a few of these specimens appeared to be hybrids, the remainder demonstrated the continued survival of the species *C. rufus.*

Since many of the specimens assigned to *C. r. gregoryi* may indeed represent the red wolf in unmodified form, they should be considered in any effort to ascertain the relationship between *C. rufus* and *C. lupus.* But since some of the specimens may actually be the result of hybridization between *C. rufus* and *C. latrans,* their inclusion with unmodified *C. rufus* in statistical compilation could result in a misunderstanding of the situation. There is no completely objective method of separating red wolves from red wolf-coyote hybrids, but earlier work (McCarley, 1962;

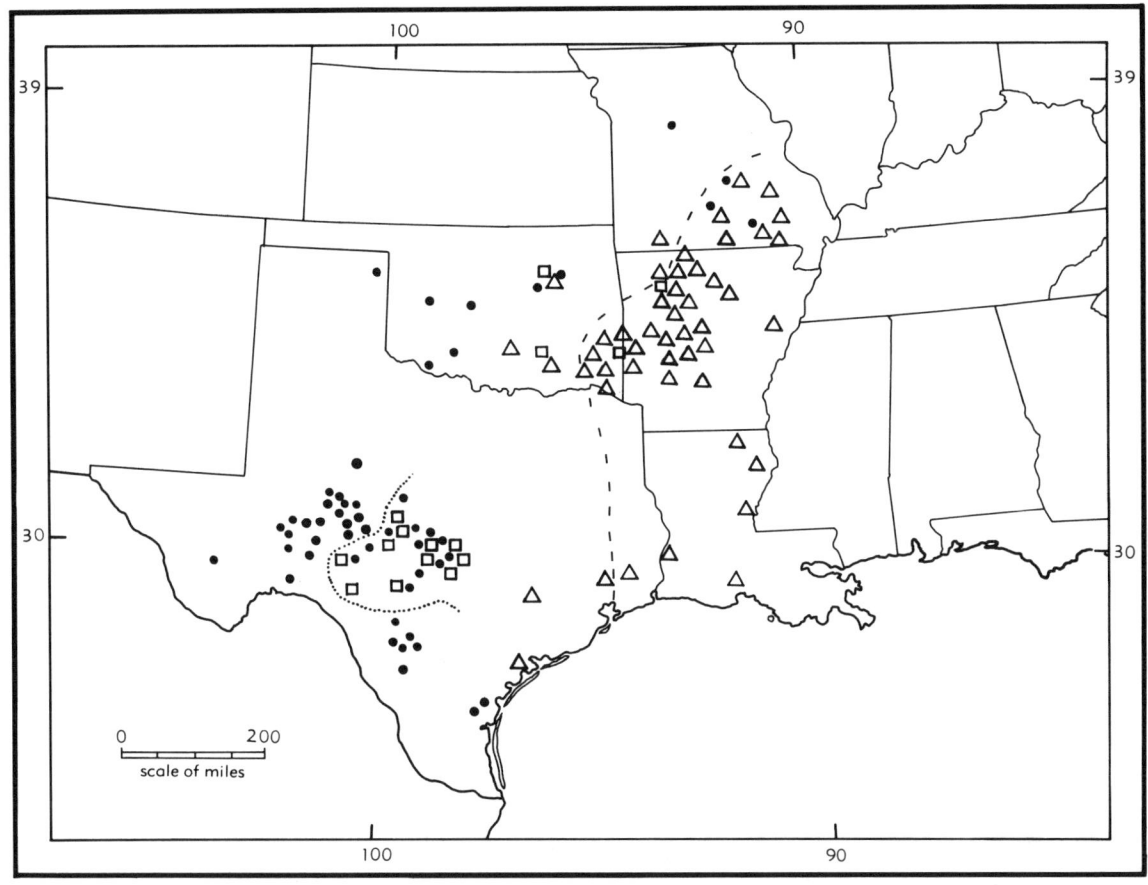

Fig. 14.—Map showing localities of specimens taken prior to 1930 in the south-central United States. Triangles, *C. rufus;* black dots, specimens originally identified as *C. latrans;* squares, specimens originally identified as *C. rufus,* but considered in this paper to represent hybridization between red wolf and coyote. See figures 18-23 for more details. The dotted line shows the western limits of oak forest in central Texas. The dashed line divides localities of specimens previously identified as *C. rufus gregoryi* (east) from those previously identified as *C. rufus rufus* (west). Note: because of the scale of the map it was not possible to plot all localities in crowded areas.

Paradiso, 1968) indicated that skulls of *C. r. gregoryi* almost invariably have a greater maximum length than those of *C. latrans.* In my preliminary examination of the series identified as *C. rufus gregoryi,* I noticed 15 adult skulls from Arkansas and eastern Oklahoma that seemed comparatively short (less than 215 millimeters in greatest length for males, and less than 210 millimeters for females), and which were within or near the size range of my standard series of western coyotes. These 15 specimens were considered the most likely to represent hybridization,

and hence were not directly compared to the standard series of *C. lupus.* They are dealt with later in this account.

Of the remaining specimens collected from 1919 to 1929 that Goldman assigned to *gregoryi,* those that could be subjected to multivariate analysis were 63(1) males and 52(1) females (see Fig. 14; and appendix A, part 10). Of these 115 skulls, 64 were from Arkansas, 19 were from southern Missouri, 29 were from southeastern Oklahoma, and three were taken in northern or western Louisiana in 1928 (a specimen collected in

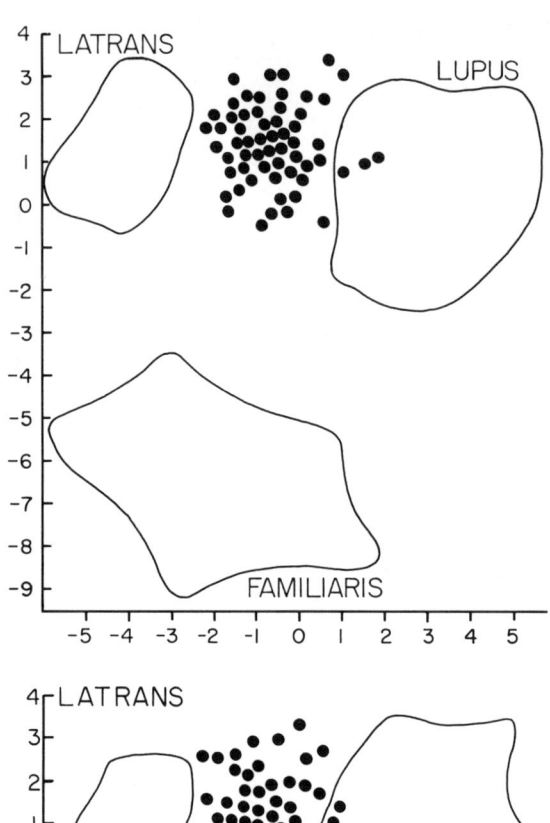

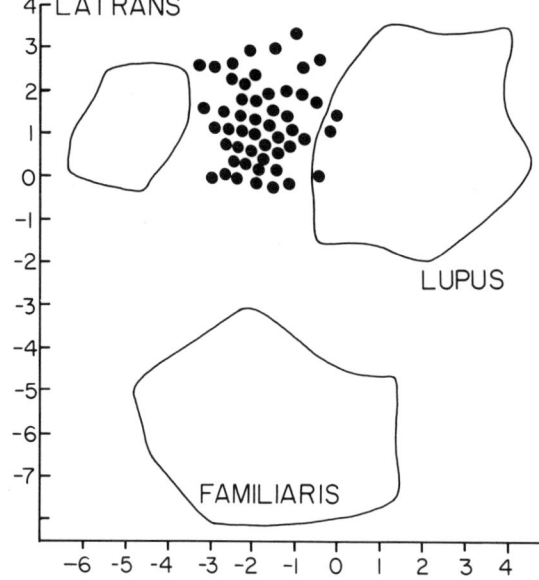

Fig. 15.—Multivariate positions (black dots) of individual specimens of *C. rufus* taken from 1919 to 1929 in the south-central United States, relative to the ranges of variation of the total series of *C. lupus, C. latrans,* and *C. familiaris* shown in Fig. 8. Males above, females below.

TABLE 3

Statistical distance, D^2, between *C. lupus, C. latrans,* and *C. rufus.*

	C. lupus	*C. latrans*
A. Males		
C. latrans	71.9	
C. rufus	18.6	22.2
B. Females		
C. latrans	63.4	
C. rufus	16.9	18.8

These 115 skulls were compared individually, by multivariate analysis, to the standard series of *C. lupus, C. latrans,* and *C. familiaris.* Their distribution (Fig. 15), like that of the earlier material from the southeast (Fig. 11), is intermediate to the graphical limits of *C. latrans* and *C. lupus.* There is no overlap with the coyote sample, but three male and three female specimens, all from the Ozark region of Arkansas and Missouri, fall within the range of variation of the gray wolf. Therefore, on the basis of a multivariate analysis involving 15 cranial measurements, complete separation of *C. lupus* and *C. rufus* is not possible. The amount of overlap between the two, however, is much less than that found among the various subspecies of *C. lupus.*

Measurements of these 115 *gregoryi* are listed in appendix B (part 5). Means of males are compared with those of other series in a ratio diagram (Fig. 13). Size and proportion in both the early red wolf and the later series of *gregoryi* are seen to match closely, and to differ from *C. lupus* and *C. latrans.* There is no indication that the later and more northerly concentrated series of *gregoryi* is substantially different or any more coyotelike than the early Louisiana material. These two series thus were combined into a single group that was considered as a standard sample of *C. rufus.* The D^2 statistics comparing the three total standard samples of *C. lupus, C. latrans,* and *C. rufus,* were calculated and are shown in table 3. The intermediate val-

extreme southern Louisiana in 1919 was included with the early eastern material covered above; see pp. 28-30).

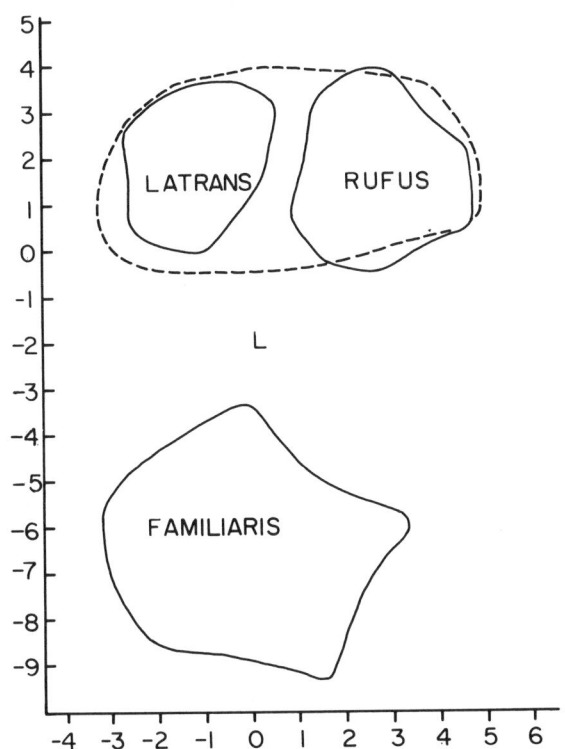
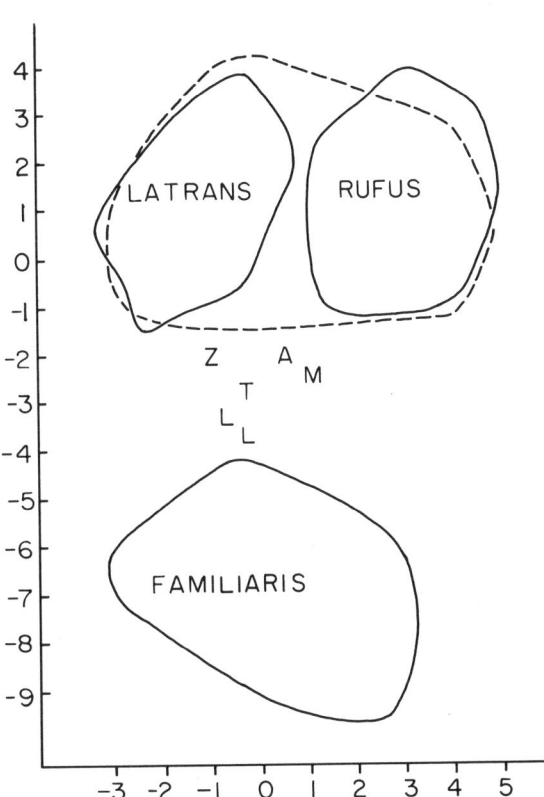

Fig. 16.—Multivariate comparison of 886 individual specimens of *Canis* from the south-central United States, to the ranges of variation of the total series of pre-1930 *C. rufus gregoryi*, the total series of *C. latrans* from northern and western North America, and *C. familiaris*. The dashed lines show the margins of the statistical distribution of 879 of the individual specimens. The seven exceptions are from Arkansas (A); Louisiana (L); Missouri (M); Lavaca County, Texas (T); and Van Zandt County, Texas (Z). See appendix A (part 11) for additional details. Males are right, females left.

ues of the red wolf are almost as near to those of the coyote as to those of the gray wolf.

Influence of the Domestic Dog on Wild *Canis* of the Southeast

Since it has been concluded that *C. rufus* is specifically distinct from *C. lupus,* and since there is no evidence that *C. lupus* existed in the southeastern United States in the twentieth century, the gray wolf need not be considered further in discussion of the systematic problems of this region. Henceforth, this section of the paper deals primarily with the relationships between *C. rufus* and *C. latrans*, and with their comparative status in the southeast. Before proceeding, however, the question of interbreeding between *C. familiaris* and wild *Canis* should be covered.

In order to evaluate the influence of the domestic dog, the three previously established samples of *C. familiaris*, *C. latrans*, and *C. rufus* were compared as groups with 886 other individual skulls collected in the southeastern and south-central states. All of these specimens had been previously identified in collections as wild *Canis,* and they include the members of nearly all of the samples referred to in the following account of the southeastern situation. The graphical results of the analyses are shown in figure 16. There is no statistical blending of wild southeastern

Canis and *C. familiaris*, and no suggestion of massive introgression from the domestic dog into any of the wild populations. Only seven specimens have statistical positions or D^2 values indicating hybridization involving *C. familiaris* (see appendix A, part 11).

Pre-1930 Relationship of Red Wolf and Coyote in Arkansas, Missouri, and Oklahoma

With the gray wolf and domestic dog eliminated from the picture, we now are dealing exclusively with the relationship of the red wolf and coyote. As an initial step, standard series of both these species were compared with each other by multivariate analysis. For the red wolf, these series consisted of the previously used 70(3) males and 55(2) females collected before 1930 and identified as *C. rufus gregoryi*. For the coyote, the previously established standard sample of western material was used, except that one male specimen was withheld for consideration at a later point. The comparative series of *C. latrans* thus consisted of 165(2) males and 111(4) females. The results of these analyses, demonstrating clear separation between the two species, are depicted in figure 17.

All remaining group comparisons involving problems in the southeast are based on the statistical distribution of *C. rufus* and *C. latrans* shown in figure 17. All other specimens, mostly those collected farther to the west or later in time than the standard sample of *C. rufus*, are evaluated on the basis of their relative distance from this sample and from the standard sample of *C. latrans*. There is no objective manner of setting definite limits to the ranges of variation shown by red wolves, coyotes, and hybrids between the two. It would therefore be meaningless to attempt to assign each and every specimen to one of these three categories. Therefore I decided to examine each group of specimens, collected in a given area and over a certain

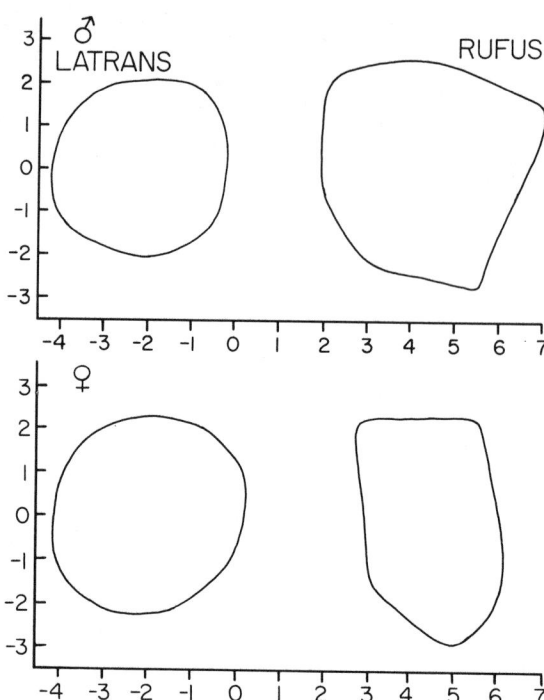

Fig. 17.—Multivariate comparison of the series of pre-1930 *C. rufus gregoryi* and the series of *C. latrans* from northern and western North America. Only the margins of the range of variation of each species are shown. Males are above, females below; *C. rufus* to the right, *C. latrans* to the left. All subsequent graphs of multivariate analyses that involve the southeastern or south-central United States are based on the illustrations shown here. The numbers along the vertical and horizontal axes are canonical coordinates. The position of a single female *C. rufus* is not included within the range of variation shown for that species. The coordinates of this specimen fall beyond the coverage of the computer plot, though not in the direction of the range of variation of *C. latrans*. The specimen is large and might possibly represent a mislabeled male. The position of this specimen is among those plotted for females in Fig. 15.

period, and to try to interpret relative affinity on the basis of the total group position.

We have already seen that a large group of specimens identified as *C. rufus gregoryi* may be clearly distinguished from a large series of western coyotes. The great majority of these specimens referred to *gregoryi* were collected between 1919 and 1929 in the Ozark-Ouachita uplands of Arkansas, southern Missouri, and eastern Oklahoma. Coyotes also were found along the western and northern edges of this region, but the original limits of their range are unknown. According to Audubon and Bachman (1851:152), *C. latrans* was "well known throughout the western part of the States of Arkansas and Missouri."

The coyotes of this region are called *C. latrans frustror*. This name was given by Woodhouse (1851) to a canid from the Cimarron River, about 100 miles west of Fort Gibson, Oklahoma. For a while, there was uncertainty regarding whether *frustror* applied to the coyote or to the larger wolf group. Merriam (1897) considered the holotype to be a coyote, but Bailey (1905:175) subsequently noted: "A series of topotypes of *frustror* secured since at Red Fork, Ind. T., shows it to be a widely different species, more nearly related to *Canis rufus*." Bailey seems not to have considered the possibility that both red wolf and coyote could have occurred together in the same vicinity. Jackson (1951:271) considered *frustror* a subspecies of *C. latrans* with a range from eastern Kansas and Missouri south to the Texas Gulf coast.

Also, according to Jackson (1951:274): "A very few specimens from the Ozark region of Arkansas and Missouri superficially hint that there may be possible hybridization, but probably not intergradation, with the Mississippi Valley red wolf, *Canis niger gregoryi* Goldman, in that region." And Lawrence and Bossert (1967) found a series of eight skulls, collected in 1921 at Fallsville, Newton County, Arkansas, and identified as *C. rufus gregoryi*, to "span the whole range of variation from coyote to wolf."

As stated above, I did not include 15 skulls, identified as *gregoryi*, in my standard sample of *C. rufus*, because they seemed unusually small in greatest length. These 3(1) males and 12(1) females were all obtained in Arkansas and eastern Oklahoma between 1919 and 1929 (see appendix A, part 12).

They were compared individually to the standard red wolf and coyote series, and their relative statistical positions are plotted in figure 18. Twelve of the specimens cluster within or near the range of variation of *C. rufus*. It is impossible to say whether these specimens are small *gregoryi* or represent genetic influence from *C. latrans*. But the fact that their positions are concentrated close to *C. rufus*, and do not form an evenly distributed bridge between red wolf and coyote, suggests the former possibility. Although these 12 skulls are short, compared to most *C. rufus*, multivariate evaluation of all measurements indicates affinity to this species. Of the other three specimens, one taken at Fallsville, Newton County, Arkansas, occupies

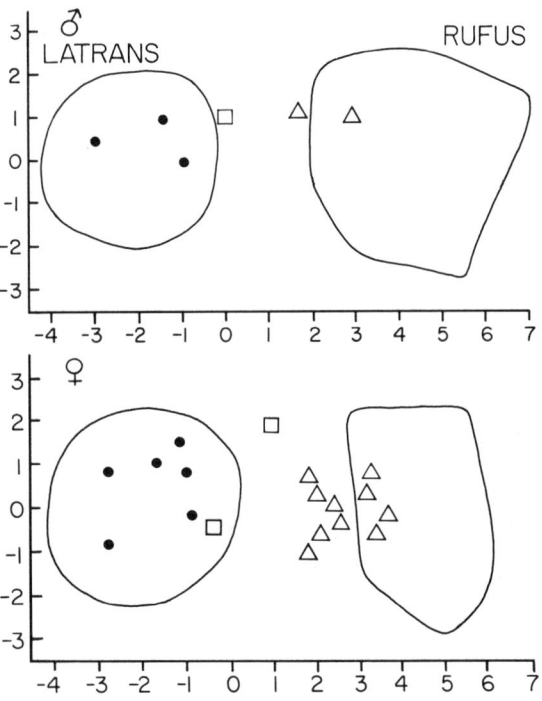

Fig. 18.—Multivariate positions of certain individual specimens relative to the ranges of variation of the series of *C. rufus* and *C. latrans* shown in Fig. 17. Black dots, *C. latrans frustror* taken prior to 1930 in Missouri; triangles and squares, specimens originally identified as *C. rufus gregoryi* that have unusually short greatest lengths. The squares indicate specimens that probably represent *C. latrans* or red wolf-coyote hybrids.

such an intermediate position (D^2 from coyote 23.4, from red wolf 24.7) that it almost certainly represents a hybrid. The two remaining specimens appear to be coyotes.

All of the questionable specimens were obtained in Arkansas and eastern Oklahoma. In southeastern Texas, Louisiana, and states farther east, the only specimens of wild *Canis* taken prior to 1930 are clearly referable to *C. rufus*. In Missouri, however, in addition to the 19 specimens that were identified as *C. rufus gregoryi* and used in my standard sample of red wolves, nine skulls were collected from 1923 to 1925 and originally identified as *C. latrans frustror* (see appendix A, part 13). They were taken in the same period and in approximately the same area of the southeastern part of the state (see Fig. 14) as the series of *gregoryi*. The multivariate positions of these specimens all fall within the range of variation of the standard coyote sample (Fig. 18). Therefore, as first pointed out by Paradiso and Nowak (1972a), there is direct evidence that *C. rufus* and *C. latrans* occurred sympatrically in this area, without intergradation or hybridization.

Until now, this section of the paper has dealt only with red wolves identified as *C. r. gregoryi*, and not with *C. r. rufus*. Most of the pre-1930 specimens that Goldman assigned to the latter named subspecies were collected in Texas, but a few were taken in Oklahoma and northwestern Arkansas. In addition, Hall and Kelson (1952:340-341) assigned two specimens taken in 1923 and 1924 at Reeds Spring, Stone County, southwestern Missouri, to *C. r. rufus*, although Goldman had listed them as *C. r. gregoryi*. McCarley (1962) suggested that the taxon *C. r. rufus* might actually represent the result of hybridization between *C. r. gregoryi* and *C. latrans*, but Paradiso and Nowak (1972a) continued to recognize its validity.

Since specimens assigned to *C. r. rufus* are on the whole smaller and more narrowly proportioned, and hence more coyotelike than most specimens of *gregoryi*, there was no need herein to depict a direct comparison of *C. r. rufus* with the standard sample of *C. lupus*. But one skull (USNM 8098), obtained sometime in the nineteenth century at Cherokee Town, in what is now Garvin County, central Oklahoma, within the designated range of *C. r. rufus*, is of special interest. It was placed in the National Museum's collection of *C. lupus nubilus*, and was not included by Goldman (1944) in his list of specimens examined. Nevertheless he did handwrite "*C. rufus*" on the specimen tag. My comparison of this skull with standard series of *C. lupus*, *C. latrans*, and *C. familiaris* shows that its position is close to that of other old southeastern material identified as *C. rufus* (Fig. 11).

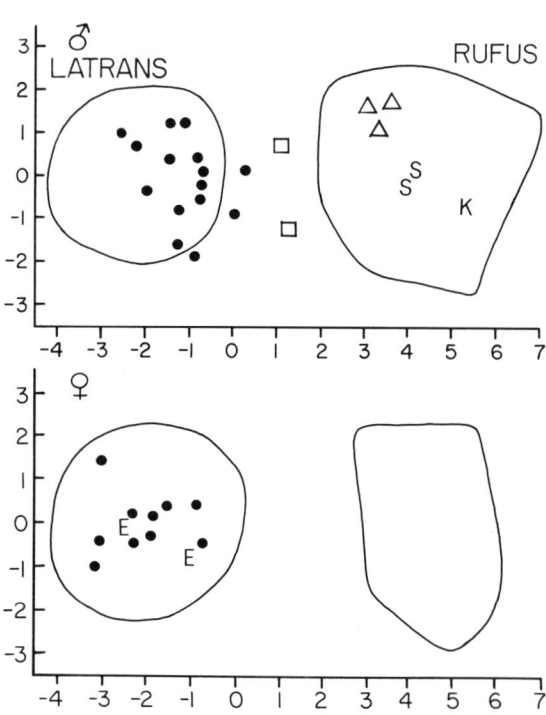

Fig. 19.—Multivariate positions of certain individual specimens relative to the ranges of variation of the series of *C. rufus* and *C. latrans* shown in Fig. 17. Squares and triangles, specimens from eastern Oklahoma and Arkansas originally identified as *C. rufus rufus*; S, specimens from Reeds Spring, Stone County, Missouri; K, specimen from Garvin County, Oklahoma; black dots, *C. latrans* from western and central Oklahoma; E, *C. latrans* from eastern Oklahoma.

Figure 19 depicts the multivariate positions of eight males identified as *C. r. rufus* (including USNM 8098). Only males of *rufus* were available from Arkansas, Oklahoma, and Missouri (see appendix A, part 14). Six fall well within the limits of *C. rufus*, whereas two of the Oklahoma skulls have intermediate positions (D^2 from *C. rufus* 28.8 and 31.4, from *C. latrans* 26.8 and 31.3). Although available data are thus inconclusive regarding the status of *C. r. rufus* in this region, some of the animals obtained here are indistinguishable from red wolves taken farther east. A number of skulls identified as *C. latrans* also were taken prior to 1930 in Oklahoma, mostly in the western and central parts of the state, and in an adjoining county of Texas (see Fig. 14; appendix A, part 15). Figure 19 shows the relative statistical positions of 15(7) males and 12(6) females to fall mainly within the limits of the standard sample of western *C. latrans*. Two females from Creek and Tulsa counties, northeastern Oklahoma, are also included within these limits. Interestingly, of a series of three specimens taken at Red Fork, Tulsa County, Oklahoma, in 1904-1905, one is statistically identical to the coyote, one is identical to the red wolf, and one is intermediate in characters.

Although there is evidence of early hybridization between *C. rufus* and *C. latrans* at certain localities in Arkansas and eastern Oklahoma, genetic exchange appears to have remained very limited before 1930. Nearly all specimens taken until then in the lower Mississippi Valley, and identified as *C. rufus*, are statistically separate from western coyotes and from coyotes taken in the south-central United States. Hence the sum of available evidence indicates that the species *C. rufus*, in essentially unmodified form, survived through the 1920's in this region. But by that period the species was under heavy pressure from man and already was making its last stand in the Ozark-Ouachita uplands. Before proceeding with a discussion of the red wolf's decline, however, it is necessary first to describe the complex pre-1930 situation in Texas.

History of Texas *Canis*

As we have seen, Audubon and Bachman (1851) wrote that the "Red Texan Wolf" occurred from northern Arkansas through Texas and into Mexico, although the only specific place they mentioned was 15 miles west of Austin, where Goldman (1944:488) fixed the type locality for *C. r. rufus*. Other early naturalists reported on the presence of wolves in this same general area, and, like Audubon and Bachman, observed considerable local variation in color. For example, Roemer (1849:80), referring to an 1846 visit to a plantation in what is now Colorado County, Texas, wrote that "the owner of the farm offered several wolf skins for sale. He had taken them from wolves recently caught in steel traps. The pelts were of various colors, one black, the other yellow, and still another greyish brown. The farmer informed us that such variation in color was quite common among the larger wolves. They were very plentiful in the forest surrounding his house and a number of hogs had been killed by them."

References are few as to how far east in Texas the coyote originally occurred, but there is no evidence of its former presence beyond the prairies. One early source (Fisher, 1841:33) discussing Brazoria County in 1840, stated: "The large black wolf abounds in the country; but the small prairie wolf of the western states, I think, is seldom, if ever, found so far south."

Bailey's detailed biological survey of Texas (1905:171-177) included the following references to *Canis* in the state.

C. lupus.—Still common over most of the plains and mountain country of western Texas, mainly west of the one hundredth meridian.

C. latrans.—More or less common over at least middle and southern Texas and apparently eastward on strips of prairie as far as Gainesville [Cooke County] and Richmond [Fort Bend County]. There are vague reports of a small wolf occurring farther east on the coast prairie even to the border of Louisiana, but specimens are needed before these reports can be associated with definite species. East of the semiarid mesquite region coyotes are rare and probably mere stragglers. True to their name of prairie wolf, they do not enter the timbered country to any extent, although at home in the scrub oak, juniper, mesquite, and chaparral, as well as over the open prairies of the southern part of the state.

C. rufus.—A definite range can be assigned the species, covering the whole of southern Texas north to the mouth of the Pecos and the mouth of the Colorado, and still farther north along the strip of mesquite country east of the plains, approximately covering the semiarid part of the Lower Sonoran zone. As yet there are no specimens to show whether these wolves extend into the more arid region west of the Pecos. While apparently nowhere overlapping the range of the larger, lighter-colored 'lobo' or 'loafer' [*C. lupus*] of the plains, they take its place to the south and east as soon as the plains break down and the scrub oak and mesquite country begins, but their whole range is shared with the coyote. The ranchmen invariably distinguish between them and coyotes, and with good reason, for the wolves kill young cattle, goats, and colts with as much regularity as the coyotes kill sheep. While paying a bounty of $1 or $2 for coyotes, the ranchmen usually pay $10 or $20 for red wolves.

Although Bailey considered *C. rufus* to occupy all of southern Texas, and to share its entire range with *C. latrans*, some modification of this delineation is suggested by the Bureau of Biological Survey field reports (on file, National Fish and Wildlife Laboratories, U.S. National Museum of Natural History) upon which his published work was based. These reports (made by Bailey, J. H. Gaut, A. H. Howell, W. Lloyd, and H. C. Oberholser from 1891 to 1905) were prepared at a time when little was known about the systematics of *Canis* in southern Texas. They refer to *C. rufus* as either a large species of coyote or a wolf. But more important than the names applied is the fact that the presence of both this larger species of *Canis* and a distinct smaller species (*C. latrans*) was generally recognized. The reports suggest that the larger species, or the animal now referred to as the red wolf, was originally common only along the Texas coast above Nueces Bay, and that *C. latrans* was rare or entirely absent from this area. Farther inland, in the more arid parts of southern Texas, the red wolf progressively diminished in numbers, whereas the coyote became abundant. The reports do not provide any insight on the original situation farther north in the Edwards Plateau area.

Examination of Pre-1930 Material from South Texas

The main problem presented by the above information is the relationship between the red wolf and coyote where their ranges met along the Texas coast, and farther inland toward the Rio Grande Valley. Unfortunately, the number and distribution of available pre-1930 specimens are not sufficient for thorough analysis of the situation. Although a useful series of coyotes was collected in south Texas, few skulls were saved from the critical areas immediately to the north of Nueces Bay and inland toward San Antonio.

A group of eight skulls, previously identified as *C. rufus rufus*, was taken in 1900 and

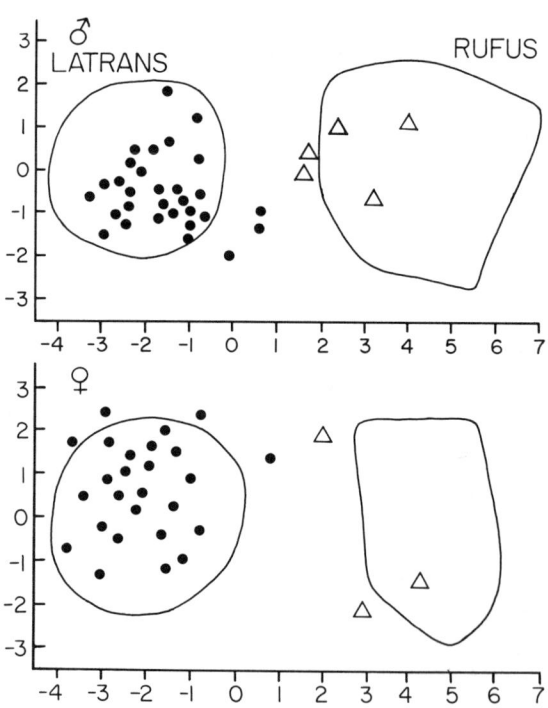

FIG. 20.—Multivariate positions of pre-1930 specimens from southern Texas relative to series of *C. rufus* and *C. latrans*. Triangles, *C. rufus rufus;* black dots, *C. latrans texensis.*

1904 near the Texas coast in Calhoun, Colorado, and Liberty counties (see appendix A, part 16). The multivariate positions of these specimens are shown in figure 20. Clearly, these specimens have close affinity with those of the standard red wolf sample, which were taken farther east and identified as *C. rufus gregoryi*, but they are smaller and more narrowly proportioned. In other words, they fit Goldman's (1944:487) description of the subspecies *C. r. rufus*. A specimen taken in 1906 in Hardin County, extreme southeastern Texas, and identified as *C. r. gregoryi*, was incorporated above in my standard red wolf sample. There are no other usable, pre-1930 specimens of red wolves obtained in southern and coastal Texas.

Skulls of 31(5) males and 26(8) females labeled as *C. latrans texensis* were taken from 1891 to 1918 in south Texas (see map, Fig. 14; appendix A, part 17), more than half coming from Nueces County. Most of these specimens fall within the range of variation of western *C. latrans* (Fig. 20). Although three Nueces County specimens have statistical positions that are removed in the direction of the standard red wolf sample, their D^2 values confirm closer affinity to *C. latrans*. Therefore, although there is suggestion of limited hybridization, early specimens from south Texas may be separated into two distinct groups, representing the species *C. latrans* and *C. rufus*.

Examination of Pre-1930 Material from Central Texas

Early specimens from the Edwards Plateau area of central Texas are not so easily divisible into recognized species, and have been the cause of much confusion. A report by Allen (1896:75-76) is pertinent, since it contains both the earliest account of possible hybridization in the area, and an early reference to the presence there of three kinds of *Canis*. Allen's informant, H. P. Attwater, said that *Canis lupus* was "formerly common in Bexar County, but I have not heard of their occurrence here for several years. They are still found in the broken, hilly country northwest of San Antonio, particularly in Edwards County." And, regarding *C. latrans*, he stated: "In Kerr County and adjoining counties they are the 'thorn in the side,' of the sheepmen. Mr. Lacey [Howard Lacey, a rancher said to be a careful and reliable observer] says that the Coyotes of that region are different from the Coyotes of the prairies, being much larger. They are believed by the ranchmen to be a cross between the 'Lobo' (Wolf) and the Coyote. Two years ago, when the bounty act was in force, the regular 'Lobo' price was allowed for the large Coyotes of the rocky region to the northwestward of San Antonio."

With regard to the red wolf and coyote in central Texas, Goldman (1944:481) reported: "Specimens collected in the vicinity of Llano, Tex., include typical examples of both species and individuals not sharply distinctive of either. Close approach in essential details and the apparent absence of any invariable unit character suggests the possibility of hybridism in some localities in Texas." McCarley (1962) considered a series of eight skulls from the vicinity of Llano to suggest the occurrence of hybridization. Paradiso and Nowak (1972a) concluded that specimens taken throughout the Edwards Plateau area of Texas between 1890 and 1918 represented a hybrid swarm of *C. rufus* x *C. latrans*. For purposes of this paper, all available early material from central Texas, as well as that collected in western Texas in the same period, was re-examined and subjected to more critical statistical evaluation.

To facilitate an interpretation of the situation, the central and west Texas material was divided on an arbitrary geographic basis. Figure 21 shows multivariate positions of specimens collected in Texas counties to the west and just north of Tom Green County; figure 22 does the same for skulls taken in Tom Green County; and figure 23 shows positions of specimens taken in the following counties of central Texas: Blanco, Burnet, Coleman,

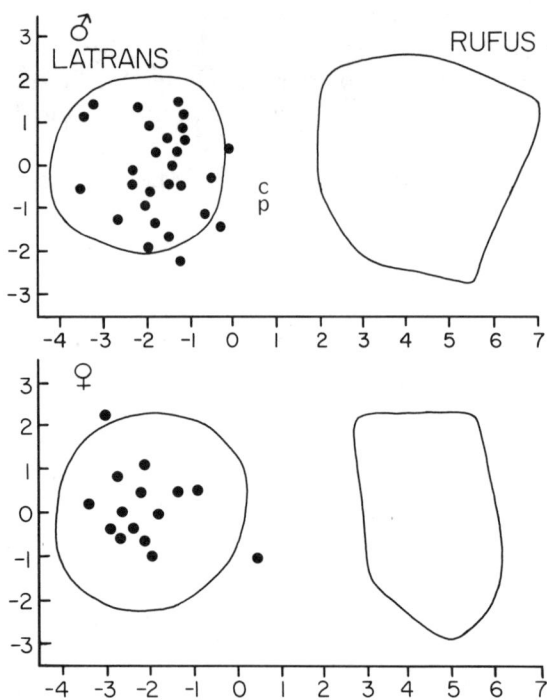

Fig. 21.—Multivariate positions of certain individual specimens relative to series of *C. rufus* and *C. latrans*. Black dots, *C. latrans texensis* from western Texas; P, specimen from Pecos County, Texas originally identified by Goldman (1944) as *C. rufus rufus*, but considered in this paper to probably represent *C. latrans texensis*; C, specimen of *C. latrans lestes* from Conejos County, Colorado.

Concho, Edwards, Gillespie, Kerr, Llano, McCulloch, Menard, San Saba, and Sutton (see also Fig. 14; appendix A, parts 18-20). All material was collected from 1915 to 1918, except for the specimens from Edwards, Gillespie, and Kerr counties, which were taken from 1899 to 1906.

Some 30 males and 15 females from western Texas, each previously identified as *C. latrans texensis*, fall within or near the range of variation of the standard coyote sample. The skulls of 36 males and 14 females taken farther east in Tom Green County, each also previously identified as *texensis*, show the same kind of statistical distribution. Hence there appears to be no clinal shift, in the direction of *C. rufus*, as the range of *C. latrans* passes from the western states, across western Texas, and into Tom Green County.

Goldman (1944) assigned to *C. r. rufus* a male from 22 miles north of Sheffield, Pecos County, which is about 100 miles farther west than the closest other record of *C. rufus* listed by him. Paradiso and Nowak (1972a) considered the skull to represent a coyote and to hardly be distinguishable from some specimens of *C. latrans lestes*. While making a preliminary examination of my standard coyote sample, I noticed one large skull of *lestes* from Bountiful, Conejos County, southern Colorado, that closely resembled the Pecos County specimen. The Colorado skull was withheld from the coyote series in the multivariate analysis comparing the standard samples of *C. rufus* and *C. latrans*. Its relative position is plotted in figure 21 and demonstrates that a few western coyotes are statistically well removed from normal *C. latrans* in the direction of *C. rufus*. The Pecos

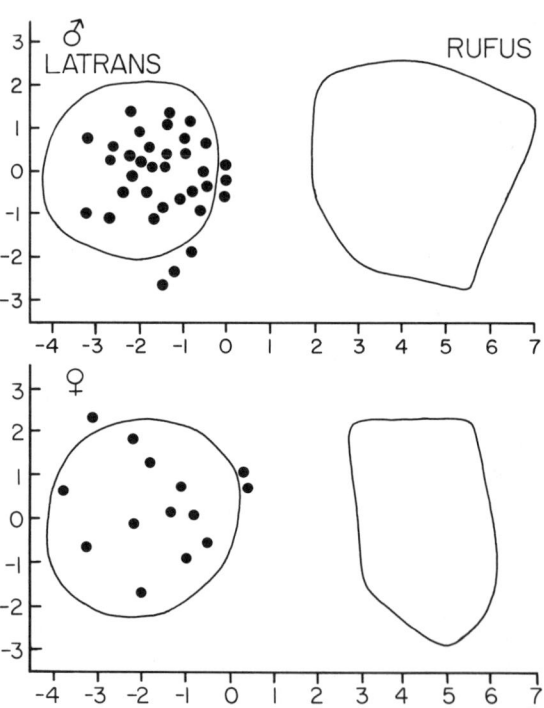

Fig. 22.—Multivariate positions (black dots) of individual pre-1930 specimens of *C. latrans texensis* from Tom Green County, Texas, relative to series of *C. rufus* and *C. latrans*.

County specimen, the position of which is close to that of the Colorado skull, could be such a coyote. The D^2 values of both skulls indicate closer affinity to *C. latrans* than to *C. rufus*, but there is no way of definitely determining whether the Pecos County specimen represents some genetic influence from the red wolf. The skin of this specimen was examined, but could not be distinguished from those of other coyotes taken in central and western Texas.

Central Texas is a land of major natural transition. Here the high plains break down into the rough country of the Edwards Plateau; the arid Sonoran area of the Lower Austral Life-zone passes into the humid Austroriparian area; and typical prairie and desert vegetation merge with eastern deciduous forest. Regarding the transition of the vegetation pattern on the Edwards Plateau, Bray (1904:14) noted that "conditions vary enough to give in some places, as in well-watered and sheltered canyons, a relatively luxuriant growth, while in other situations, as upon stony arid slopes, there is the scantiest vegetation."

At the turn of the century, central Texas also was undergoing pronounced changes because of the influence of man. Overgrazing by cattle and sheep, and the frequent occurrence of fire, had suppressed the native prairies and permitted the spread of mesquite. Simultaneously, clearing and cultivation along the river bottoms had deprived the area of its richest forest growth (Bray, 1904). And, of course, the hand of man was turned heavily against predatory animals such as wolves and coyotes.

The specimens of *Canis* collected in central Texas were originally identified as *C. rufus rufus*, *C. latrans texensis*, and *C. latrans frustror*. Jackson (1951:271, 279) considered this last named subspecies to range north of Nueces Bay, and to the eastern edge of the Edwards Plateau where he recorded specimens from Blanco, Burnet, Llano, and San Saba counties. Jackson wrote that *texensis* occurred to the south of Nueces Bay, but that its range extended northwestward into central Texas (he listed specimens there from Coleman, Concho, Gillespie, Kerr, McCulloch, Menard, and Sutton counties). Goldman (1944:488-489) thought that *C. rufus rufus* occurred throughout central Texas, and he recorded pre-1930 specimens from Burnet, Edwards, Kerr, Llano, and McCulloch counties.

Those specimens from central Texas that could be subjected to multivariate analysis included animals identified as the following: *C. latrans texensis*, male—13, female—12(2); *C. latrans frustror*, male—9, female—15; *C. rufus rufus*, male—21(1), female—5. Localities are plotted on the map in figure 14, and statistical positions are shown in figure 23. Also shown are positions of the 14 skulls of *C. lupus monstrabilis* collected in central and western Texas, each of which was used above in my standard gray wolf sample (most were collected from 1900 to 1920). Their positions are plotted here to emphasize that, although *C. lupus* was present in the area, it was distinct from the animals being called red wolves, and was not a factor in the problems under discussion.

The multivariate positions of the central Texas skulls identified as *texensis*, *frustror*, and *rufus* form a statistical bridge between the ranges of variation of standard *C. latrans* and *C. rufus*. There is no meaningful place to draw a line separating these skulls into coyotes and red wolves, and we apparently are dealing with some sort of blending between the two. Specimens collected in certain restricted localities, such as those from the vicinity of Llano and Burnet, by themselves bridge the statistical gap between red wolves and coyotes. After an evaluation of these statistics, a visual re-examination of the skulls, and a check of available skins from central Texas, it must be concluded that the original identification of these specimens was arbitrary and based on an incomplete understanding of the situation. In this regard it is

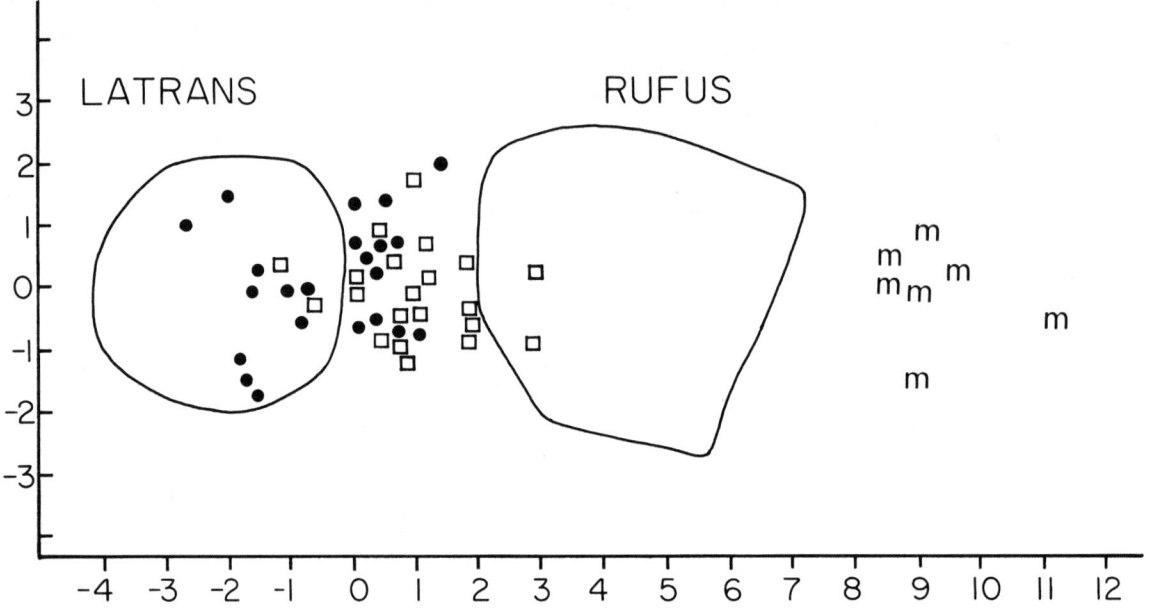
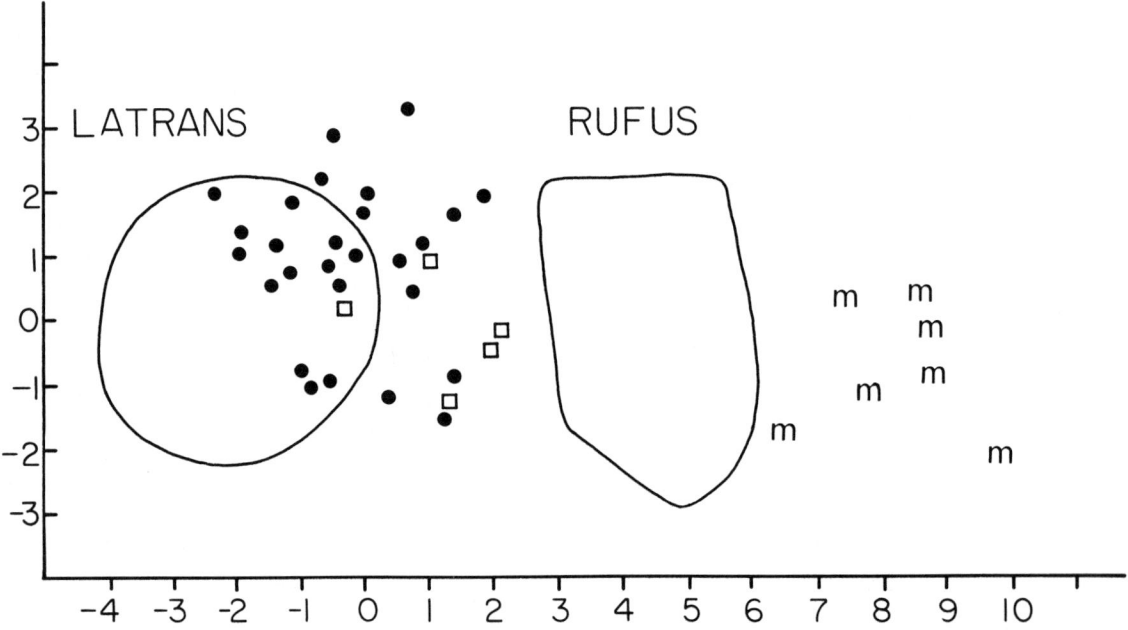

FIG. 23.—Multivariate positions of certain individual pre-1930 specimens from Texas, relative to series of *C. rufus* and *C. latrans*. Black dots, specimens from central Texas originally identified as *C. latrans texensis* or *C. latrans frustror;* squares, specimens from central Texas originally identified as *C. rufus rufus;* M, specimens of *C. lupus monstrabilis* from central and western Texas. Males are above, females below.

interesting to note an unusual sex ratio among the available specimens of *rufus* and *frustror* from the area. For specimens originally identified as *frustror*, females outnumber males 15 to 9; but for skulls previously referred to *rufus*, males outnumber females 20 to 5. This suggests to me that the original identification of the specimens was based partly on the bias of assigning larger skulls, predominantly males, to *rufus*, and smaller skulls, predominantly females, to *frustror*.

No other group of specimens, at least among those collected before 1930, falls to such an extent between the statistical limits of *C. latrans* and *C. rufus*. This condition could have resulted from one of the following factors: (1) intergradation in central Texas between the small western coyote and the larger eastern red wolf; (2) long-term interbreeding, and the production of a hybrid zone serving as a bridge for the flow of genes from one species to the other; (3) short-term interbreeding caused possibly by drastic alteration of the environment.

I do not think that the statistical distribution of specimens is indicative of intergradation and hence the conspecificity of red wolf and coyote. Recognized subspecies of *C. latrans* intergrade throughout western North America and show no morphological overlap with red wolves. There is no evidence of a cline of characters approaching those of standard *C. rufus*, even among specimens taken as far east as Tom Green County, Texas. The sudden breakdown found in the limited area of central Texas is highly atypical of North American *Canis*, and does not represent normal subspecific intergradation. Furthermore, as we have seen, there is no suggestion of original intergradation in any other area where the ranges of *C. rufus* and *C. latrans* met, and there is direct evidence of the sympatric occurrence of the two species in Missouri (see pp. 35-36).

While I therefore consider hybridization responsible for the situation in central Texas, there is no reason to believe that the process had been occurring over a long period. Had interbreeding been going on for many years or centuries, its effects seemingly would have spread beyond central Texas.

Available evidence favors the theory that hybridization between *C. rufus* and *C. latrans* was of relatively short-term occurrence, and probably had begun in the latter part of the nineteenth century in response to man's disruption of the habitat and his persecution of native wild canid populations. This is in accord with earlier statements by McCarley (1962) and Paradiso and Nowak (1972a). Since central Texas was an area of natural transition, where the eastern red wolf and western coyote would have overlapped in range, if not in habitat, opportunities for interbreeding undoubtedly occurred. Once man altered the environment, and at the same time attacked the wolves and coyotes, ecological and behavioral isolation might have broken down, and large-scale hybridization begun.

According to Mayr (1963:128): "By far the most frequent cause of hybridization in animals is the breakdown of habitat barriers, mostly as a result of human interference." Mayr (1963:118-121) used the term "hybrid swarm" to describe populations, in such areas of breakdown and interbreeding, that form a continuous bridge between two parent species. This term seems applicable to the population of *Canis* in central Texas.

The exact original limits of the ranges of *C. rufus* and *C. latrans* in central Texas are unknown, partly because the hybridization factor has obscured the picture. The red wolf, however, usually is considered a species of the eastern forests. In this regard it is interesting to note that the area of central Texas under discussion falls just within the extreme western boundary of eastern forest elements (see Fig. 14). Indeed, oak trees first appear immediately to the east and south of Tom Green County.

At this point I would like to summarize briefly the early status of *Canis* in the southeast, as evaluated on the basis of specimens collected prior to 1930. The red wolf, a species distinct from the gray wolf of northern and western North America, was found along the Atlantic coast and westward probably to the edge of the prairies. Large series of specimens show that before 1930 the red wolf in the lower Mississippi Valley was easily separable from the western coyote. The two species occurred sympatrically, or at least showed no tendency to intergrade in most areas where their ranges approached. A small percentage of specimens taken in eastern Oklahoma and Arkansas suggest that limited hybridization had occurred at certain localities in those states. In central Texas, however, more extensive interbreeding had resulted in the formation of a hybrid swarm between the two species.

Survival of the Red Wolf from the 1930's to 1950's

Although the red wolf, in unmodified form, seems to have survived in the lower Mississippi Valley through the 1920's, the species had been under heavy human pressure throughout most of its range. The course of its decline was discussed by me (1967, 1970, 1972, 1974), Russell and Shaw (1972), and Young (1944). Wolves disappeared from Pennsylvania, Maryland, West Virginia, Virginia, North Carolina, South Carolina, Tennessee, Kentucky, Ohio, Indiana, and Illinois without specimens having been saved which would have enabled us to determine their systematic status.

Farther to the south, wolves, probably only *C. rufus*, did not survive much longer. Harper (1927:315-317) said that the last known kill of *C. r. floridanus* in the Okefinokee Swamp of Georgia took place about 1908, although there were later reports. There seem to be no other definite records of wolves in Georgia after 1900. According to Chapman (1894:345) wolves in Florida were already "on the verge of extinction." The last reported occurrence in the state, listed by Young (1944:25), was in 1903 near the Everglades. H. H. Bailey (1930) wrote that a wolf was killed about 1918 or 1920 on the Osceola side of the Kissimee River.

Howell (1921:30) reported that wolves in Alabama were "on the verge of extinction. Their last stronghold appears to be the rough, hilly country stretching from Walker County northwestward to Colbert County." The last recorded kill took place south of Cherokee, Colbert County in 1917. Apparently, however, wolves held out for a while longer in the state. In 1937 and 1938 reports by the Predator and Rodent Control branch of the U.S. Bureau of Biological Survey (on file at the offices of the Division of Animal Damage Control, U.S. Fish and Wildlife Service, Washington, D.C.) it is said that in Alabama nine wolves were killed in 1937 and three in 1938. Holliman (1963:242) reported a specimen of *C. rufus* collected at Livingston, Sumter County, western Alabama in 1944. The multivariate position of its skull falls within the statistical limits of the standard red wolf sample (Fig. 24).

Wolves apparently disappeared from most of the higher country of Mississippi at an early date, but survived until comparatively recently along the Mississippi River and Gulf Coast. Jenkins (1933:125-127) reported "but few wolves in Mississippi from 1892 to the present time." He added that during the great flood of 1927 many wolves crossed into southwestern Mississippi from Louisiana. According to Young (1944:29), these wolves were still present in 1932 and were preying on livestock. Goldman (1944:485) reported, but did not examine, a specimen of *C. rufus gregoryi* taken at Biloxi in 1931. I found this skull to fall squarely within the range of variation of standard *C. rufus* (Fig. 24).

Very little information is available concerning the possible former occurrence of the red wolf in Kansas. Lantz (1905) referred to "*Canis ater*" as "Once abundant. Still found in a few scattered sections of the state." Cockrum (1952:229) cited reports of

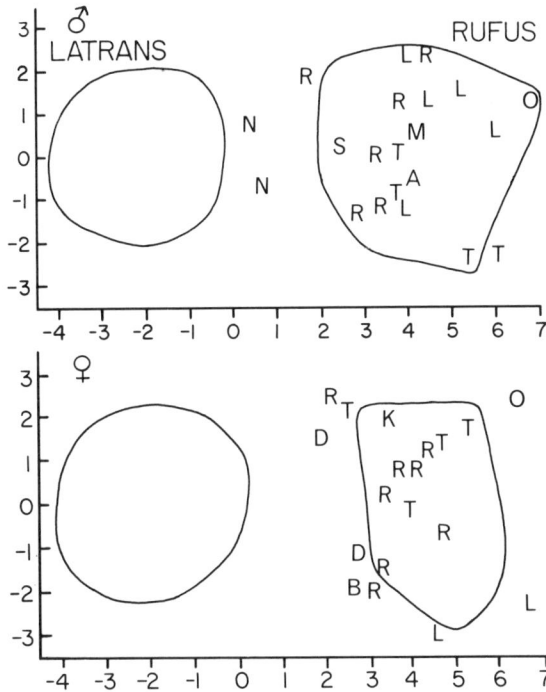

Fig. 24.—Multivariate positions of specimens of *C. rufus*, and two hybrids, taken from the 1930's to the 1950's. A, Alabama *gregoryi*; K, Arkansas *gregoryi*; L, Louisiana *gregoryi*; M, Mississippi *gregoryi*; O, Oklahoma *gregoryi*; T, Texas *gregoryi*; R, Texas *rufus*; B, specimen of *C. r. rufus* taken in 1954 in Brazoria County, Texas; D, specimens of *C. r. rufus* taken in Madison County, Texas (originally identified as *C. latrans*, but considered in this paper to represent *C. rufus*); S, specimen of *C. r. rufus* taken in southern Harris County, Texas (originally identified as *C. latrans*, but considered in this paper to represent *C. rufus*); N, specimens of probable red wolf-coyote hybrids taken in northern Harris County, Texas.

a red wolf killed in Cherokee County, extreme southeastern Kansas, in 1908 or 1909, and of one heard howling there in 1915.

The situation in the Ozark-Ouachita uplands, and adjacent areas, is most confusing in the 1930's and 1940's, and is discussed in detail in the next subsection of this paper. I will first deal with those areas in which the red wolf seems to have survived past 1930 in unmodified form.

Six skulls obtained in northern Louisiana from 1935 to 1940 are indicative of the continued presence of *C. rufus* in that state (Fig. 24). No other specimens of wild *Canis* taken between 1930 and 1950 are available from Louisiana. All evidence suggests that up until the last few years of this period the red wolf maintained moderate numbers over much of Louisiana. By the early 1950's, however, the species had been decimated in most areas, except for the eastern bottom lands and southern marshes of the state (Nowak, 1967). McCarley (1962) reported specimens of *C. rufus* obtained in 1956 and 1957 in Terrebonne and Madison parishes, respectively. The skull taken in Terrebonne Parish, southern Louisiana, falls within the statistical limits of the red wolf (Fig. 24). The skull from Madison Parish, northeastern Louisiana, was not suitable for multivariate analysis, but does represent *C. rufus*.

According to Gipson (1972:4), red wolves were present in southern Arkansas as late as the 1950's. A specimen taken in 1942 on the Union-Columbia county line, just north of the Louisiana border, has a multivariate position within the limits of *C. rufus* (Fig. 24).

A male and a female specimen taken in 1936 near Battiest, McCurtain County, extreme southeastern Oklahoma were reported by McCarley (1962) to be red wolves. These two skulls are statistically well removed from my standard coyote sample (Fig. 24), and probably indicate continued survival of *C. rufus* in the area.

The southeastern part of Texas, especially the area known as the Big Thicket, seems to have been one of the last major refuges for the red wolf. Goldman (1944:486-489) recorded post-1930 specimens of *C. rufus gregoryi* from Hardin, Newton, and Polk counties; and *C. r. rufus* from Brazoria, Brazos, Liberty, Montgomery, and Walker counties. In addition, Jackson (1951:275) listed specimens of *C. latrans frustror* from Harris County, in the same area. Most of this material, plus two other skulls in the National Museum from Madison County that had been mistakenly labeled as *C. latrans texensis*, were subjected to multivariate analysis. The results (Fig. 24) confirm the survival of the red wolf in southeastern Texas. As in other cases, specimens identified as *C. r. rufus* tend to show some statistical approach to *C. latrans*. The two Madison County specimens apparently represent *C. r. rufus*. One of the Harris County skulls was collected at Genoa on the south side of Houston. Although originally identified as *C. latrans frustror*, it falls within the range of variation of the standard red wolf sample, and I think it actually represents *C. rufus*. The other two Harris County skulls were taken at Humble, on the north side of Houston, and their intermediate statistical positions suggest genetic influence of the coyote at that locality. All of the above listed Texas specimens were collected between 1930 and 1943, and I examined very little material taken over the next 20 years in the state. The skull of a female taken in eastern Brazoria County in 1954 has a multivariate position suggesting the survival of *C. r. rufus* in the area.

In summary, it appears that the red wolf continued to exist in parts of its former range from the 1930's to the 1950's. No usable specimens of wild *Canis*, other than those mentioned in this subsection of the paper, were

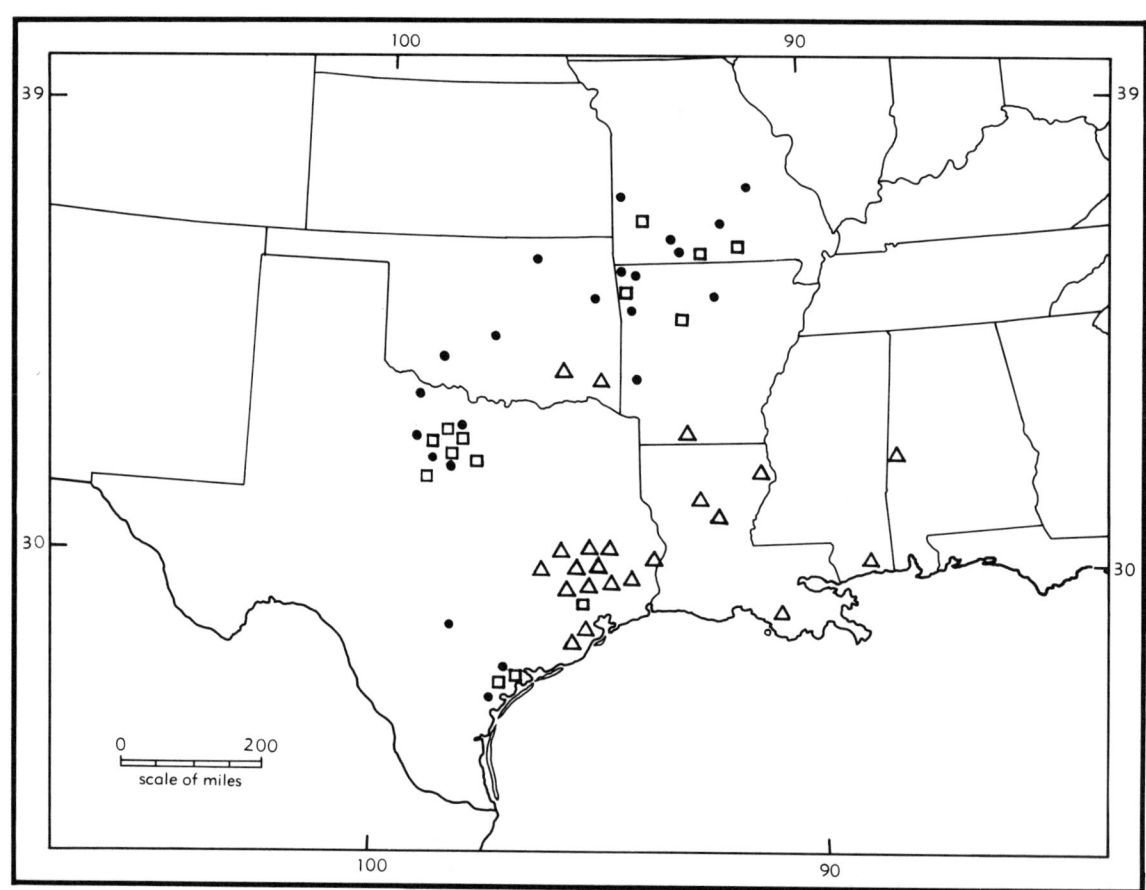

Fig. 25.—Map showing localities of specimens taken from the 1930's to the 1950's in the south-central United States. Triangles, *C. rufus;* black dots, specimens originally identified as *C. latrans;* squares, specimens originally identified as *C. rufus,* but considered in this paper to represent hybridization between red wolf and coyote (also two specimens taken in northern Harris County, Texas originally identified as *C. latrans*). See Figs. 24, 26, 27, 28, 29, and 30 for more details. Note: because of the scale of the map it was not possible to plot all localities in crowded areas.

collected in the areas under discussion. Although the distribution of the red wolf was certainly not stable through the entire period, it seems safe to say that about 1940 the range of unmodified *C. rufus* extended from extreme southeastern Oklahoma and southeastern Texas, across Louisiana, southern Arkansas, and southern Mississippi, and into western Alabama. Localities of the available specimens from this region are listed in appendix A (part 21), and are plotted in figure 25.

Increase of Hybridization from the 1930's to 1950's

Specimens collected prior to 1930 indicate the effects of hybridization between *C. rufus* and *C. latrans* only in central Texas, and at a few localities in Arkansas and eastern Oklahoma. In the following decades, specimens of apparent mixed genetic origin were taken over larger areas. Information from the 1930's and early 1940's seems important to a full understanding of the phenomena then engulfing populations of southern *Canis*, but, unfortunately, comparatively few specimens were saved from the areas of greatest interest.

As explained earlier, specimens taken in 1900 in Calhoun County, on the middle Gulf Coast of Texas, have multivariate positions within or near the limits of standard *C. rufus*. No other specimens of wild *Canis* were collected in this area until after 1930. Jackson (1951:275) listed specimens of *C. latrans frustror* from Aransas, Refugio, and Victoria counties. Goldman (1944:488-489) stated that specimens of *C. rufus rufus* had been collected in the same area. Skulls of 10 males obtained from 1936 to 1942 in these three adjoining counties, were suitable for analysis (see Fig. 25; appendix A, part 22). Five had been referred originally to *C. rufus* and five to *C. latrans*. As shown by the positions plotted in figure 26 these previous designations are questionable and some genetic exchange seems to have occurred.

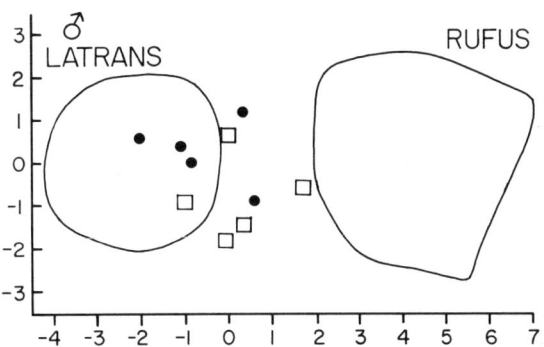

Fig. 26.—Multivariate positions of individuals taken from 1936 to 1942 in Aransas, Refugio, and Victoria counties on the central coast of Texas. Squares, specimens originally identified as *C. rufus*; black dots, specimens originally identified as *C. latrans*.

Few specimens from farther inland in southern and central Texas are available from this period. No skulls at all were saved from the central Texas counties that were apparently occupied by a hybrid swarm of *C. rufus* x *C. latrans* prior to 1920 (see pp. 41-43). Probably, as is shown in the next subsection of the paper, all wild *Canis* had been exterminated in this area by the 1930's. Three males of *C. l. texensis*, taken in 1942 in Bexar County, have multivariate positions within the range of *C. latrans* (Fig. 27).

Farther north, a series of specimens was taken from 1930 to 1942 in the area between the Colorado and Red rivers (see Fig. 25; appendix A, part 23). Records of both *C. rufus* and *C. latrans* were listed from various localities in this area by Goldman (1944) and Jackson (1951), but I can see no basis for separating these specimens into two species. I subjected 18 of the skulls to multivariate analysis and found most to fall within or near the range of variation of standard *C. latrans* (Fig. 27). The more intermediate positions of two females (D^2 from coyote 36.7 and 22.8; from red wolf 39.3 and 25.6) suggest genetic influence from *C. rufus*.

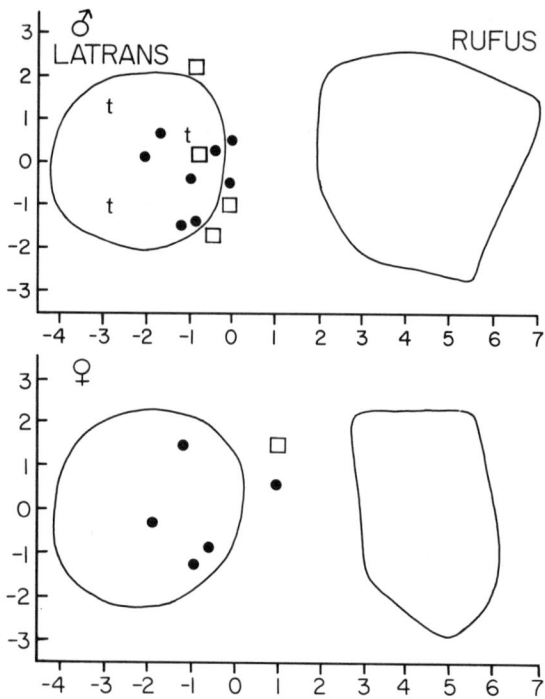

Fig. 27.—Multivariate positions of individuals taken in Bexar County, Texas (t), and in north-central Texas from 1930 to 1942. Black dots, specimens originally identified as *C. latrans*; squares, specimens originally identified as *C. rufus*.

From 1933 to 1942 a large series of skulls was collected on the Wichita Mountains National Wildlife Refuge, Comanche County, southwestern Oklahoma (see Fig. 25; appendix A, part 24). Nearly all of these specimens were originally identified as *C. latrans frustror*, and this designation is supported by statistical positions of 22 males and 25 females (Fig. 28). Goldman (1944:488) listed a single specimen of a male from this group as *C. rufus rufus*. I, however, was not able to distinguish this skull from those of some large coyotes, and its multivariate position (Fig. 28) suggests affinity with *C. latrans*. Therefore, if the red wolf ever did occur as far west in Oklahoma as Comanche County, it apparently disappeared before the 1930's. Nonetheless, the presence of coyotes, some of which had perhaps received an introgression of genes from *C. rufus*, caused persons to think that wolves still inhabited the area. According to Halloran and Glass (1959:363): "The Texas red wolf reaches the western edge of its range in the Wichita Mountains. Most of the canids present on the refuge are coyotes, but the situation is clouded by the local custom of calling everything larger than a small coyote a wolf." And Duck and Fletcher (1945:128) reported that *C. rufus rufus* was being taken over most of the state.

Several other specimens, identified as *C. l. frustror*, were taken in central and northeastern Oklahoma in 1932. Those usable in

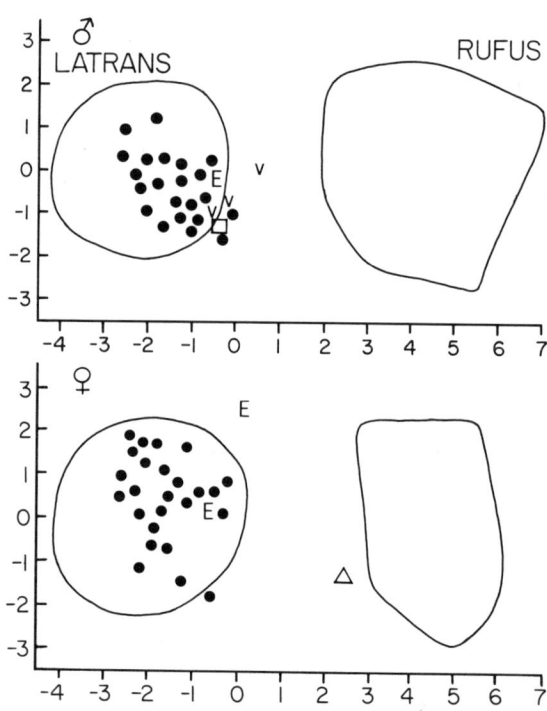

Fig. 28.—Multivariate positions of individuals taken in Oklahoma from 1932 to 1942, relative to series of *C. rufus* and *C. latrans*. Black dots, *C. latrans frustror* from Comanche County; square, specimen from Comanche County identified by Goldman (1944) as *C. rufus*, but considered in this paper to represent *C. latrans*; v, specimens from Cleveland County; E, coyotes from eastern Oklahoma; triangle, specimen from Atoka County apparently representing *C. r. rufus*.

multivariate analysis include three males from Cleveland County, one male from Cherokee County, and two females from Osage County (see appendix A, part 25). These six specimens demonstrate statistical approach to *C. rufus* (Fig. 28). A single female, taken the same year in Atoka County, southeastern Oklahoma, and identified originally as *C. r. rufus*, seems to be within the range of variation of other specimens assigned to that subspecies (Fig. 28). Two other skulls (in USNM) collected in 1932 in eastern Oklahoma are a female from Cherokee County identified as *C. r. rufus*, and a male from LeFlore County assigned by Goldman to *C. r. gregoryi*. Neither one is suitable for multivariate analysis, but both appear to be intermediate in characters between the standard samples of *C. rufus* and *C. latrans*. As mentioned in the last subsection of this paper, two skulls of red wolves were collected in 1936 in McCurtain County, southeastern Oklahoma.

The distribution of specimens thus suggests that by the 1930's the red wolf had become restricted to parts of southeastern Oklahoma. A few skulls taken in the northeastern and central areas of the state, however, may represent the results of interbreeding between *C. rufus* and *C. latrans* (see Fig. 25).

Specimens discussed earlier in the paper demonstrate the sympatric occurrence of *C. rufus* and *C. latrans* in southern Missouri prior to 1930. By that year, however, the red wolf had been decimated by government and private hunting (Nowak, 1970; Sampson, 1961). In contrast, the coyote seems to have maintained its numbers and increased its range. One possible indication of subsequent hybridization in the area was provided by Bennitt and Nagel (1937:168) who wrote: "It is difficult to outline exactly the range of these two species in Missouri, since so many observers cannot tell them apart."

According to Sampson (1961), the last pure specimen of a Missouri red wolf was collected in 1932 in Dade County. This specimen, however, was not listed by either Goldman (1944) or Jackson (1951), and its multivariate position (Fig. 29) is well within the range of variation of *C. latrans*. A specimen collected the same year in Iron County was listed by Goldman (1944:486) as *gregoryi*, and was considered by Paradiso and Nowak (1972a:11) possibly to be a red wolf. At present, however, I think the skull represents hybridization involving *C. familiaris* (see Fig. 16).

Sampson (1961) reported that in 1941 and 1942 a series of 171 specimens was taken throughout Missouri, partly in order to de-

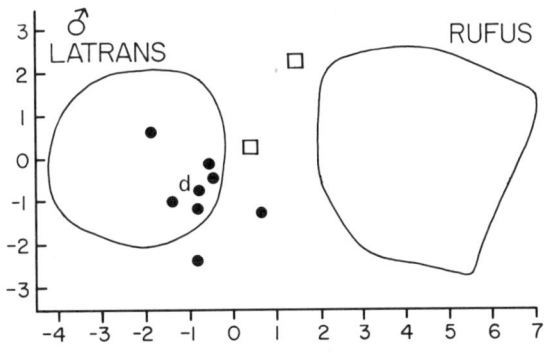
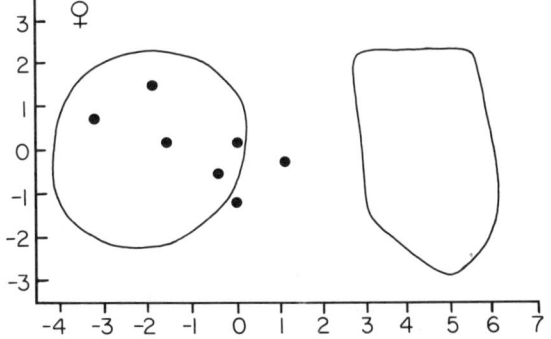

Fig. 29.—Multivariate positions of individuals taken in Missouri from 1932 to 1942, relative to series of *C. rufus* and *C. latrans*. Black dots, specimens originally identified as *C. latrans*; d, specimen from Dade County listed as *C. rufus* by Sampson (1961); squares, specimens from Ozark and Oregon counties identified as *C. r. rufus* by Leopold and Hall (1945).

termine whether the red wolf was still present in the state. He wrote that 44 of the skulls were sent to the U.S. National Museum, but suggested that these included all specimens thought most likely to represent *C. rufus*. Hence the sample sent in probably already was biased in favor of larger and more wolf-like individuals. In any case, the Museum reportedly identified the specimens as 39 coyotes, one dog, one coy-dog, and one red wolf-coyote hybrid. Jackson (1951:274) referred most of these specimens to *C. latrans frustror*, and listed records all across southern Missouri. Of the specimens taken in 1941 and 1942 in the southern half of Missouri, eight males and seven females were suitable for multivariate analysis (see Fig. 25; appendix A, part 26). Their relative positions are shown in figure 29, and indicate close affinity with *C. latrans*, but possibly some genetic influence from *C. rufus*.

Elder and Hayden (1977) evaluated 20 specimens of *Canis* from Missouri, deposited in the collection of the University of Missouri, and reported one to be *C. familiaris* and two to be coy-dogs. Of the others, four taken between 1945 and 1950 in the Ozarks were identified as *C. rufus*, nine taken mostly from 1940 to 1955 were identified as *C. latrans*, and four taken between 1946 and 1949 were considered to probably be red wolf-coyote hybrids. The graphical results of this analysis suggest to me that these last 17 specimens represent a single interbreeding population, and not that the red wolf survived as a distinct entity in Missouri until 1950. The fact that Elder and Hayden also identified the two skulls mentioned in the next sentence as *C. rufus*, suggests that their accepted limits for this species were less restrictive than my own.

Leopold and Hall (1945) referred to *C. r. rufus* two specimens of males, taken in 1941 and 1942, respectively, in Ozark and Oregon counties, extreme southern Missouri. In a 1952 letter cited by Sampson (1961), Leopold wrote: "In all respects these two specimens are intermediate between typical coyotes and typical red wolf." Multivariate analysis of the skulls (Fig. 29) supports this statement. Leopold and Hall had assigned the specimens to *C. r. rufus* on the basis of their small size, even though they were obtained within the range designated for *C. r. gregoryi* by Goldman (1944:484-486). Since Goldman had assigned two male specimens from Reeds Spring, Stone County, Missouri to *gregoryi*, an awkward situation developed in which material of *rufus* was reported from farther to the east. Hall and Kelson (1952:340-341) technically solved this problem by stating that the two Stone County skulls resembled *C. r. rufus* in small size and cranial characters, and by referring them to this subspecies. The issue, however, is a temporal as well as a geographic one. The Stone County material was taken in 1923 and 1924, and as we have seen, falls right in the statistical center of a group of 70 skulls of *C. r. gregoryi* (Fig. 19). The Ozark and Oregon County specimens were taken almost 20 years later at a time when the red wolf had been nearly or completely exterminated in Missouri. They are statistically beyond standard *C. rufus*, and should, I think, be looked upon as representing hybridization between *C. rufus* and *C. latrans*, rather than a particular subspecies of the red wolf.

A related problem seems to exist just to the south in Arkansas. Locality records from this state listed by Goldman (1944:485-488) include 42 for *C. rufus gregoryi* and 6 for *C. r. rufus*. If the adjacent marginal records for each subspecies are plotted on a map, as they were by Hall and Kelson (1959:852), an awkward picture emerges. Several of the localities from which *rufus* was reported are seen to be nearly surrounded by the designated range of *gregoryi*. There are no geographic barriers that might account for such a distribution, but once again the time element seems to be a factor. As best as I can determine, all but one of the specimens assigned by Goldman to *gregoryi* were collected prior to 1930. In contrast, there is only one locality from which specimens assigned to *rufus* were taken before 1930. This locality is Boxley, Newton County, which is not one of those surrounded by the range of *gregoryi*. A skull taken in 1922 at this site falls within the standard range of variation of *C. rufus* (Fig. 19). The localities that are encircled by the range of *gregoryi* are Raspberry and Hector, both in Pope County. Of the four specimens recorded by Goldman from these places, two are not usable in my analyses, one has been identified by me as a dog hybrid (Fig. 16), and the statistical position of one is shown in figure 30. Of the other post-1930 specimens listed by Goldman, only one could be subjected to analysis, and its position is also depicted in figure 30. The specimens evaluated in the analysis, as well as the others, suggest a more pronounced influence by *C. latrans* after 1930.

Jackson (1951:274) recorded *C. latrans frustror* from a number of localities in western and northern Arkansas. Most of the specimens involved were collected between 1932 and 1942, but not all are suitable for analysis. The positions of those that were statistically evaluated are plotted in figure 30. A few other skulls, taken as late as 1951 in the same areas, are also depicted. Arkansas specimens taken in this period are listed in appendix A (part 27), and localities are shown in figure 25. The number of specimens is hardly enough to allow a full understanding of the situation, but their distribution suggests that after 1930 *C. rufus* was no longer prevalent in Arkansas, and that animals of more intermediate or coyotelike characters were beginning to predominate.

The Empty Zone and Expansion of Modified *Canis*

The maps in figures 14 and 25 show many areas in which no records of *Canis* are plotted. Of course, material from states west of Texas, and north of Oklahoma and Missouri, was available, but was not considered applicable to the problems under discussion. The records that are shown seem to form a roughly circular pattern around a blank area centered in northeastern Texas. It could be argued that this area is not represented only because no one ever collected there, or because no material from there was preserved. But it should be understood that almost all specimens discussed in this section of the paper were taken in the course of Federal predator control efforts. This control work was done in areas where complaints existed, and thus is indicative of the presence and abundance of wild *Canis*. To be sure, only a small percentage of the animals killed were sent as specimens to the National Museum, and collecting was not continued on a regular geographic or temporal basis. Nonetheless, the available material may serve as one relia-

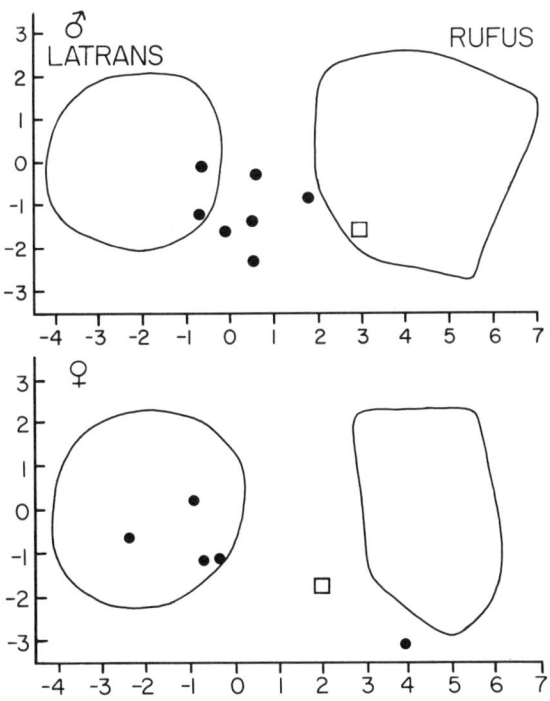

Fig. 30.—Multivariate positions of individuals taken in Arkansas from 1932 to 1951, relative to series of *C. rufus* and *C. latrans*. Black dots, specimens originally identified as *C. latrans*; squares, specimens originally identified as *C. rufus*.

ble means of evaluating the situation over a lengthy period. Therefore, the constant lack of specimens from the area in question leads me to think that no significant population of wild *Canis* had survived there into the twentieth century.

Bailey (1905:172) stated: "The black wolf is reported from a few localities in the timbered region of eastern Texas, but in most cases as 'common years ago, now very rare or extinct.'" But in another publication, perhaps because it was written for different purposes, Bailey (1907:13) stated: "In the timbered region of eastern Texas, especially in the extensive swamps and bottoms, the black wolf is still abundant and very destructive to cattle and hogs, while it renders sheep raising practically impossible." Bailey's own field reports from 1904 (on file, National Fish and Wildlife Laboratory, U.S. National Museum of Natural History) do indicate that in the area of the Big Thicket in Hardin and Liberty counties, wolves were still common. Strecker (1926:16) reported wolves in extreme eastern Texas to be "now almost or quite extinct." Apparently then, by the early twentieth century wolves had been wiped out in most of the higher country of eastern Texas, but continued to live in the bottom land swamps, especially in the area of the Big Thicket.

To the north of the Big Thicket, it seems as though there was actually a zone nearly or completely empty of wild *Canis*. This idea is supported by the existence of certain maps showing the distribution of *Canis* in Texas at various times. One of these, prepared by Bailey (1907:7), depicts a large area of north-central Texas as being free of the presence of wolves. The status of coyotes in the area, however, is not indicated.

In the annual report for fiscal 1931 of the Texas District of the U.S. Bureau of Biological Survey (on file, offices of U.S. Fish and Wildlife Service, San Antonio), there is a map showing the "Present Infestation of Predatory Animals." The term "predatory animals" is vague, but since coyotes and wolves were usually the main target of Federal control programs, especially in Texas, it is safe to assume that canids are being referred to. The map indicates that predators were absent in the eastern part of Texas, except along the coast, in the vicinity of the Big Thicket, and immediately south of the Red River. In a large area of central Texas, including all of those counties thought occupied by a hybrid swarm of *Canis* prior to 1920, the map shows predatory animals to be "under control" in 1931. This term may be taken to mean nearly, if not completely, exterminated. There is evidence that all wild *Canis* had been decimated in central Texas by this time. According to Gabrielson (1936:64), sheep raisers in this area sustained losses of ten percent to predators in 1915, but, because of Federal control programs, had losses of only a fraction of one percent in 1935. Several of the reports cited by Russell and Shaw (1971a) indicated that by the late 1930's coyotes and wolves had been completely exterminated in some central Texas counties.

Another map, drawn by Russell and Shaw (1971a), was based on reports made in 1940 by field personnel of the Texas Game, Fish and Oyster Commission. The range of the red wolf at this time was shown to extend all along the Gulf Coast north of Nueces Bay, to extend inland in southeastern Texas up the Trinity and Neches river basins, and to include strips along the Red, Sabine, and Sulphur rivers in extreme northeastern Texas. Most of the reports cited from northeastern and inland Texas, however, state that wolves were very rare in these areas. The only major area of abundance indicated by the reports cited was the southeastern corner of the state. In this regard there is agreement with the 1931 Biological Survey map and the available specimens (see pp. 45-46).

Although Russell and Shaw dealt primarily with *C. rufus*, reports by Texas game officials regarding the status of coyotes were also made between 1940 and 1942 (on file, offices of Texas Parks and Wildlife Department, Austin). These reports indicate the presence of coyotes only in those areas south of San Antonio, west of the Pecos River, and northwest of Fort Worth.

The sum of evidence from three different sources (specimens, Federal reports, and Texas government reports) suggests that the following situation existed in Texas in the 1930's and early 1940's. Red wolves were present in moderate numbers in the southeastern corner of the state and along the central Gulf Coast. In the latter area hybridization with coyotes was probably occurring. Coyotes themselves were common in the southern, western, and northern parts of Texas. In a large area centered in northeastern Texas, however, all kinds of wild *Canis* seem to have been rare, if not totally absent. I think that this empty zone may have been significant for two reasons: (1) it served as an actual barrier that for a time limited contact, and probably interbreeding, between *C. latrans* to the north and west, and *C. rufus* in southeast Texas, and hence may have assisted in maintaining an unmodified red wolf population in the latter area; and (2) it formed an empty niche for predators that awaited reoccupation by suitable canids.

This void apparently began to fill by the 1940's. As discussed in the previous subsection of this paper, coyotes and animals intermediate to *C. latrans* and *C. rufus* seemed by that time to be increasing in Oklahoma, southern Missouri, and Arkansas. In this period there are several references to the presence of wild *Canis* in Arkansas. Black (1936) wrote that the "timber wolf" was rare in the northwestern part of the state, and he did not mention the presence of coyotes. Dellinger and Black (1940) listed several records of coyotes, and also remarked: "Wolves are becoming rather common in the Ozarks." The Arkansas Game and Fish Commission (1951:96-99) reported that coyotes had been extending their range and could be found as far east as the central part of the state. Wolves, which also had reportedly increased, were said to be most common in the Ozark area, and second most common in southwestern Arkansas. But in this last named account, wolves and coyotes were discussed together and perhaps were being confused. Furthermore, it was stated that the Mississippi Valley wolf, *C. r. gregoryi*, had become rare and should not be totally exterminated. This implies recognition that the canids reported to be common were actually something other than the original native wolves of most of Arkansas. In contrast to this viewpoint, Sealander (1956:279) reported: "The race *gregoryi* occurs throughout the State and is quite numerous in some counties. It evidently has largely replaced the race *rufus* over its former range in Arkansas." Sealander, however, informed me (pers. comm.) that this statement had been based partly on Goldman's (1944:487) report that *rufus* had become restricted to parts of central and southern Texas. Since *rufus* was thus not supposed to exist in Arkansas, but since canids thought to be red wolves were certainly present, Sealander had to refer the latter to *gregoryi*.

These references suggest an increase in wild *Canis* in Arkansas between the 1930's and 1950's, with the animals involved believed to be wolves. None of these reports, however, was based on a thorough examination of specimens. The few specimens that are available from this period suggest that the original wolf population of Arkansas was being replaced by more coyotelike animals (Fig. 30). Later, Paradiso (1966) reported that three specimens taken in 1964 in Chicot County, extreme southeastern Arkansas, represented a range extension of *C. latrans frustror*. The first study of large series of Arkansas specimens collected since 1930 was described by Gipson (1972), and Gipson, Sealander, and Dunn (1974). On the basis of multivariate analysis of 284 adult skulls taken from 1968 to 1971, they concluded that the existing population of *Canis* in the state was predominantly coyote, though with some red wolf influence. The essentially coyotelike character of this population also is recognized in the next subsection of my paper. Therefore, the canids that were said to be common in Arkansas through the 1950's could not have been red wolves.

While true red wolves were being decimated in the Ozark-Ouachita uplands in the 1920's, coyotes probably moved in and increased in numbers, partially filling the vacant ecological niche. Surviving red wolves probably interbred with coyotes and produced some of the intermediate specimens discussed in the previous subsection. These hybrids served as a bridge for the introgression of genes from *C. rufus* into *C. latrans*. Much of the coyote population was thus modified, probably in a manner that favored its continued expansion into a woodland habitat. Of course, lumbering and agricultural practices may also have assisted the spread of *C. latrans* into the south-central states (McCarley, 1962; Gipson, 1972).

The increase of coyotes in Arkansas, and undoubtedly also in Oklahoma, was followed by their large-scale build up in the previously empty zone centered in northeastern Texas. Halloran (1959, 1960) provided records that almost completely block in northeastern Texas. Although he thought that these records depicted the occurrence of black-colored red wolves, the animals in question were certainly members of the expanding population of modified coyotes.

As Halloran explained, the existence of black animals was thought to indicate the presence of red wolves, rather than coyotes. Whereas *C. rufus* had a locally common black phase, only one record of a black coyote was known to Young (1951:52). Halloran (1958) also used records of black canids in an effort to plot the distribution of *C. rufus* in Oklahoma. In recent years, however, there has been widespread recognition of the existence of black coyotes in the south-central United States. Halloran himself (1963) reported one from Comanche County, Oklahoma. Dalquest (1968) wrote that a specimen of *C. latrans texensis* from Wilbarger County, Texas was in the black color phase. Pimlott and Joslin (1968) attempted to locate black animals in Arkansas, hoping that these would be *C. rufus*, but the four that they found were coyotes of medium size. Of the 284 Arkansas skulls analyzed by Gipson (1976), 24 were from black animals. Of these, 12 were identified as coyotes, six as coyote-dog hybrids, five as coyote-red wolf hybrids, and one as a domestic dog. Freeman (1976:14) reported that 12 of 121 Oklahoma specimens were black or very dark, and that the skulls of eight of these were identified as coyotes, two as coyote-dog hybrids, and two as coyote-red wolf hybrids. Elder and Hayden (1977) stated that of the seven Missouri specimens they considered to be red wolves or coyote-red wolf hybrids, five were black or had been associated with black animals.

All recent records of black coyotes have come from the former range of *C. rufus* (except for one reported in Michigan by Ozoga and Harger, 1966). Although it could be argued that this phenomenon is the result of preservation of a newly favorable mutation, I consider it further evidence of the recent introgression of genes from *C. rufus* into *C. latrans*.

The coyotelike population apparently increased rapidly in the 1950's. McCarley (1959) reported *C. latrans* to be common in most areas of east Texas. He said also that the red wolf was either extirpated or extremely rare in the area. McCarley (1962) reported data on 110 skulls collected since 1948 in east Texas, eastern and central Oklahoma, and Arkansas. Observing that over-all skull size, as represented by greatest length and zygomatic width, was the only consistent character separating red wolves and coyotes, he referred all but one of the specimens to *C. latrans frustror* (the exception was a skull from an unknown Arkansas locality, assigned

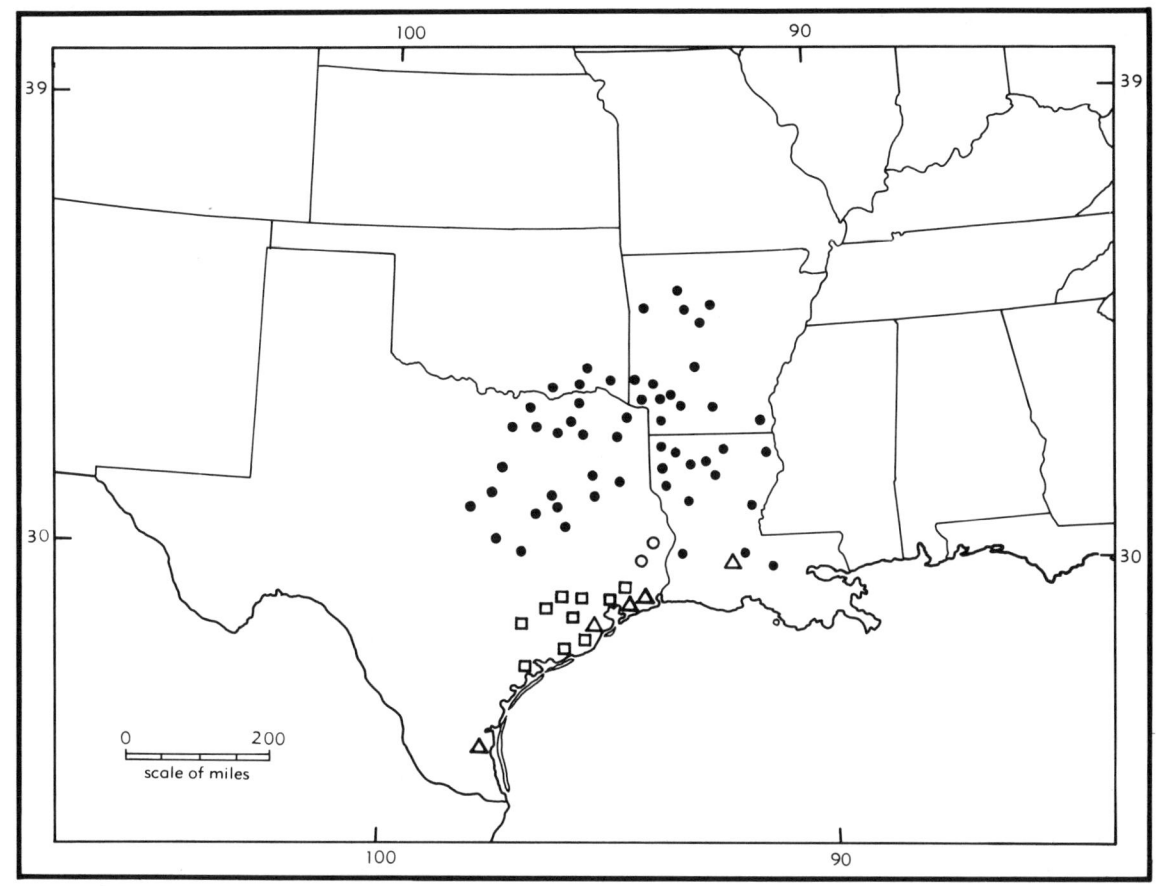

Fig. 31.—Map showing localities of specimens taken from 1961 to 1971 in the south-central United States. Triangles, *C. rufus*; squares and open circles, specimens apparently representing hybridization between *C. rufus* and *C. latrans* (see Figs. 37-39); black dots, specimens of the modified population of *C. latrans* (see Figs. 32-36). Note: because of the scale of the map, it was not possible to plot all localities in crowded areas, especially along the Texas coast.

to *C. r. rufus*). McCarley concluded that coyotes had replaced red wolves in Oklahoma, Arkansas, and east Texas, and that hybridization between the two may have occurred. He thought that *C. rufus* was extant only in a few isolated parts of eastern and southern Louisiana, on the basis of two skulls obtained there in 1956 and 1957 (see p. 45).

Evidently red wolves were present over much of Louisiana through the 1940's (Lowery, 1943; Nowak, 1967; St. Amant, 1959). By the early 1950's, however, their numbers had been greatly reduced by government trapping and private hunting. Their last major concentrations were in the bottom lands along the Mississippi River in the eastern part of the state, and in the southern coastal marshes and prairies. For a brief period, the northwestern and north-central parts of Louisiana were apparently left without a significant population of wild *Canis,* and, in effect, these areas became an extension of the "empty zone" in adjacent northeastern Texas. But just after the heaviest period of wolf trapping, from 1947 to 1952, coyotelike animals were reported to be moving into the northwestern part of Louisiana. These canids subsequently spread over most of northern and central Louisiana, and continued to increase in numbers each year (Nowak, 1967, 1970; Paradiso, 1966; Wilson, 1967).

McCarley's (1959, 1962) reports were the first published indications that the species *C. rufus* might be in serious trouble throughout its entire range. Until then, there had been a general belief that red wolves were common over large areas. Actually, Goldman (1937) had written that both the subspecies *floridanus* and *rufus* might be extinct, but subsequently (1944) he reported the latter to be present in parts of central and southern Texas. Young (1946:43) thought that red wolves could survive indefinitely in large areas of Texas and Louisiana where they would be in no direct conflict with man. Later Federal reports told of large and even increasing numbers of red wolves in Texas, Arkansas, and Oklahoma. Obviously such reports were based on the movement of coyotes into areas formerly occupied by *C. rufus*, but as late as 1964 some authorities considered the species to be in no danger (see Nowak, 1967, 1970, 1972). By that year, however, the precarious status of the red wolf was generally recognized, and efforts were underway to evaluate the threats both from man and from interbreeding.

Examination of Post-1960 Material from Oklahoma, Arkansas, Louisiana, and Inland East Texas

The disappearance of *C. rufus* from most of its former range, and its partial replacement by other canids, motivated the collecting of many new specimens. Nearly all of this material was taken in the normal course of predator control work by Federal and state employees. For convenience, the following discussion is arranged on a partly arbitrary geographic basis. Localities of all specimens discussed in this subsection (that were examined by me) are plotted on the map in figure 31, and listed in appendix A (parts 28-32).

Oklahoma.—Skulls of 12(4) males and 13(3) females, taken in 1965 in southeastern Oklahoma, all fall within or near the statistical limits of standard *C. latrans* (Fig. 32). These specimens were obtained in the same area from which a large series of *C. rufus* had been taken prior to 1930. After a multivariate analysis of 138 skulls taken in 1975 and 1976, and 114 collected prior to 1975, mostly from 1953 to 1970, Freeman (1976: 13-14, 28, 33, 47-48) concluded that the current Oklahoma population of wild *Canis* was essentially coyotelike. Of his total of 252 specimens, 203 were identified as *C. latrans*, one as *C. familiaris*, 33 as coy-dogs, and 15 as intermediate to *C. latrans* and *C. rufus*. There was no significant difference between the older and newer groups, except in the southeastern part of the state where skulls taken in 1975 and 1976 were found to be smaller. Freeman suggested, and I agree, that the genetic influence of the red wolf in eastern Oklahoma had declined to the point

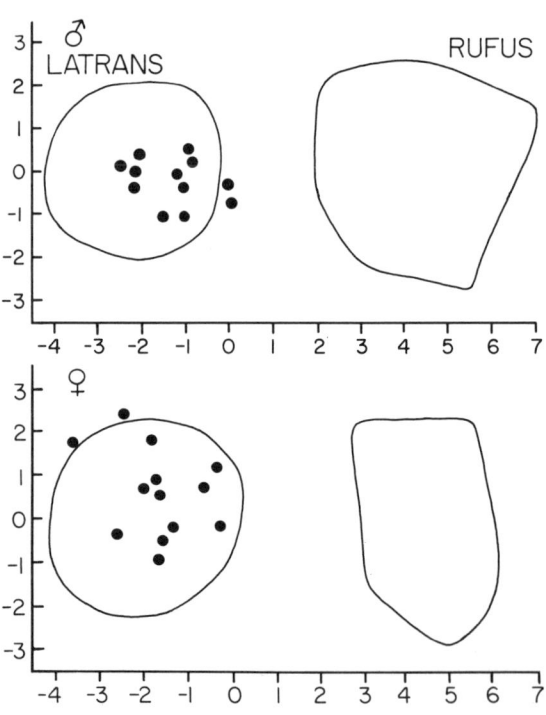

Fig. 32.—Multivariate positions (black dots) of individual post-1960 specimens from southeastern Oklahoma, relative to the ranges of variation of the series of *C. rufus* and *C. latrans* shown in Fig. 17.

at which only an occasional individual exhibited any characters of *C. rufus*.

Northern Arkansas.—Statistical positions of 20 males taken in 1969 and 1970 in the Ozark area of northern Arkansas (Conway, Franklin, Newton, Pope, and Van Buren counties), also were found to lie mainly within the range of variation of *C. latrans* (Fig. 33). Two specimens, however, have D^2 values slightly closer to *C. rufus*, thus indicating the continued genetic influence of this species. No female specimens from northern Arkansas were tested by multivariate analysis, but 20 skulls from the area were examined, and most were indistinguishable from those of western coyotes. None exceeded 200 millimeters in greatest length or 100 millimeters in zygomatic width. Gipson, Sealander, and Dunn (1974) found small "pockets of red wolf influence" in the Ozark and Ouachita mountains, but did not refer any particular specimens from those areas to *C. rufus*. Pim-

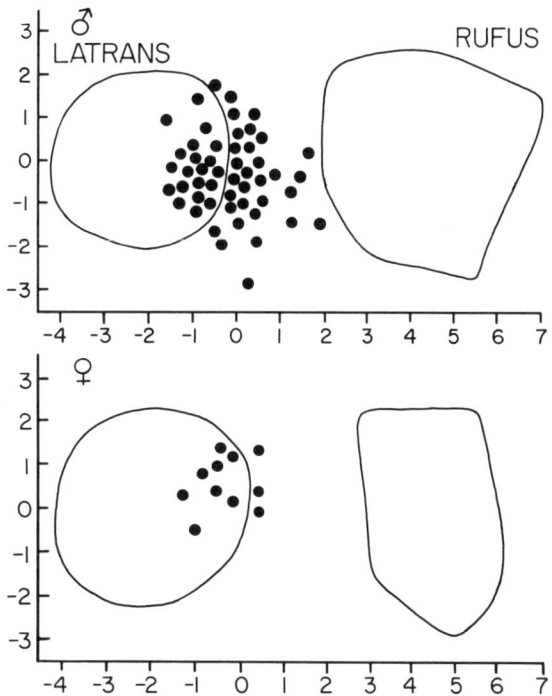

Fig. 34.—Multivariate positions (black dots) of individual post-1960 specimens from southern Arkansas, relative to series of *C. rufus* and *C. latrans*.

lott and Joslin (1968) heard what they considered probably to be wolves in the Ozark National Forest in 1964, but found only coyotes there in 1965. Subsequently, a group of investigators from Arkansas Polytechnic College continued attempts to locate red wolves in the Ozarks, both through examination of skulls and elicitation of howling responses (see Nowak, 1970). Although some wolflike calls were heard, no conclusive evidence of the presence of *C. rufus* was obtained (Henri D. Crawley, pers. comm.). There is little doubt that true red wolves have now disappeared throughout the Ozark-Ouachita uplands.

Southern Arkansas.—The multivariate positions of skulls of 52(6) males and 11(3) females, taken from 1964 to 1970 in southern Arkansas, are depicted in figure 34. Most of these fall within or near the range of variation of *C. latrans*, but there is a pronounced over-all shift toward the standard red wolf

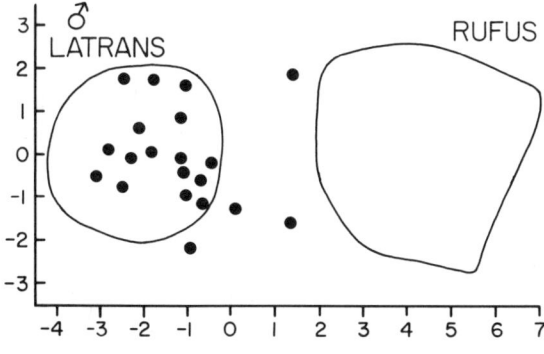

Fig. 33.—Multivariate positions (black dots) of individual post-1960 specimens from northern Arkansas, relative to series of *C. rufus* and *C. latrans*.

sample. Four individuals have D^2 values nearer to *C. rufus* than to *C. latrans*. An additional 20 skulls of females from the area, examined, but not tested by multivariate analysis, appear to follow the same pattern. Gipson, Sealander, and Dunn (1974) reported that two specimens from southern Arkansas were not significantly different from *C. rufus*, and that there was a strong genetic influence from the red wolf in the area. They added, however, that the red wolf, in pure form, probably no longer existed in Arkansas, and that its genes had been incorporated into a predominantly coyote population. Of the 284 skulls they subjected to multivariate analysis, only these two were classified as red wolves, 208 were identified as coyotes, 27 were said to be intermediate to *C. rufus* and *C. latrans*, and the remainder were considered to represent wild dogs or dog hybrids. The fact that Arkansas *Canis* is more wolflike in the southern part of the state may reflect longer survival of *C. rufus* there, and hence more recent introgression from individuals of that species. Gipson (1972:50-51) suggested that a "zone of red wolf influence" in the south-central part of the state might have resulted from the establishment of an unofficial refuge in that area, in which red wolves were released in the late 1950's.

Louisiana.—Skulls of 22(17) males and 19(17) females, taken from 1963 to 1969 in Louisiana, demonstrate similar statistical distributions to those formed by the southern Arkansas material (Fig. 35). Goertz, Fitzgerald, and Nowak (1975) concluded that 155 skulls collected in Louisiana from 1963 to 1973 represented an essentially coyotelike population that had been slightly modified through introduction of genes from *C. rufus*. Elements of this same population apparently now have spread across Mississippi and into Alabama (Cahalane, 1964; Paradiso, 1966; Linzey, 1971; Wolfe, 1972). Recent information indicates that coyotelike animals are now also established in southwestern and south-central Tennessee (D. W. Yambert, Tennessee

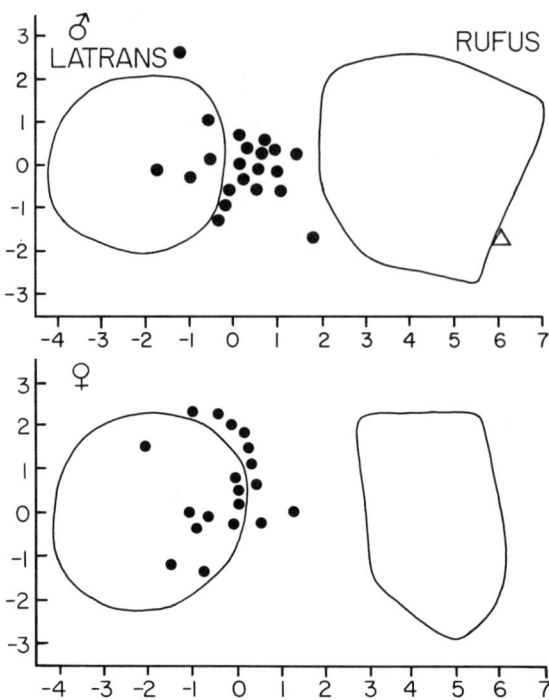

Fig. 35.—Multivariate positions (black dots) of individual post-1960 specimens from Louisiana, relative to series of *C. rufus* and *C. latrans*. The triangle shows the position of a specimen of *C. rufus gregoryi* collected in 1964 in St. Landry Parish.

Wildlife Resources Agency, pers. comm.), and in southwestern Georgia (Daniel W. Speake, Cooperative Wildlife Research Unit, Auburn University, pers. comm.). Red wolves seem to have been extirpated throughout this region, except possibly in southern Louisiana. In 1964 Pimlott and Joslin (1968) heard small groups of wolves in the northeastern part of the state and in adjoining sections of Mississippi, but there have been no subsequent records from that area. Persistent reports of red wolves have come from the coastal marshes in Cameron and Vermilion parishes, but I have examined no specimens from this area. In 1976, personnel of the U.S. Fish and Wildlife Service carried out field studies in Cameron Parish at the extreme southwestern corner of Louisiana, and live-captured several animals that appeared to be red wolves. Coyotelike individuals, however, also were reported from this area. A skull, obtained in 1964 near Washington, St. Landry Parish, south-central Louisiana, appears distinct from other recent material taken in the state. According to Pimlott and Joslin (1968:383), this specimen was identified as *C. rufus gregoryi* by Barbara Lawrence. Although the skull has several dental anomalies, its multivariate position (Fig. 35) supports this identification. Red wolves have received legal protection in Louisiana since July 1970 (Nowak, 1971).

Inland East Texas.—Figure 36 shows the statistical positions of 77(6) males and 42(6) females, collected from 1964 to 1971 in counties of east Texas more than 100 miles inland. Paradiso (1968) examined most of this material earlier, and reported that it represented an interbreeding population bridging the size gap between red wolves and coyotes. Paradiso, however, had statistically combined the inland material with a number of specimens taken along the Gulf Coast. Later, Paradiso and Nowak (1972a), also having combined inland and some coastal samples, came to about the same conclusion. They thought that the hybrid swarm, established at the turn of the century in central Texas, had expanded eastward, and by the 1960's had engulfed most of east Texas and adjacent parts of Louisiana and Arkansas. This hypothesis no longer seems fully tenable. It was pointed out above (pp. 52-53) that wild *Canis* in central Texas had been nearly extirpated by the 1930's, a period prior to the build up of *Canis* in east Texas. Thus the population of the former area could hardly have been the source for that of the latter. Furthermore, when intensively analyzed, recent east Texas material does not show exactly the same kind of statistical distribution as that of the earlier specimens. The present population is predominantly coyotelike, and does not form an even distribution bridging the gap between *C. rufus* and *C. latrans* (compare Figs. 23 and 36). Two specimens do fall within the range of variation of standard *C. rufus,* a male from Lamar County and a female from Hamilton County. There is no apparent correlation between the geographical and statistical distributions of the inland east Texas material. More wolflike and more coyotelike specimens occur throughout the area, sometimes together at the same locality (see also appendix B, part 6).

The large samples from inland east Texas, Louisiana, and southern Arkansas all have similar multivariate distributions, and apparently represent a single population with a common origin. The region occupied by this population corresponds in large part to the zone in which *Canis* was rare or absent over much of the first half of the century. Its geographical distribution also lies mainly along the southern fringe of the area in which specimens of hybrid characters appeared in the 1930's and 1940's. Such hybridization led to the introgression of *C. latrans,* and probably

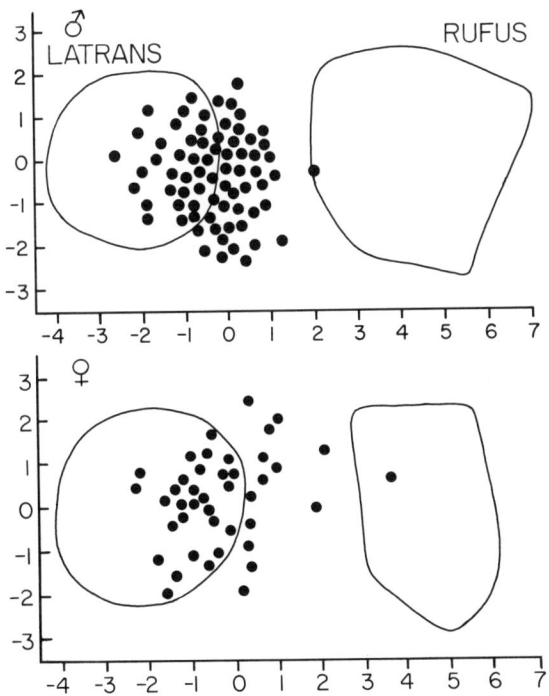

Fig. 36.—Multivariate positions (black dots) of individual post-1960 specimens from inland east Texas, relative to series of *C. rufus* and *C. latrans.*

also to the genetic swamping of any red wolves in the area that had survived man's onslaught. After having been modified through introgression from *C. rufus*, the coyote population expanded southward and eastward across the south-central states. Additional hybridization probably occurred as remnant pockets of red wolves were encountered. I do not think, however, that the current population of wild *Canis* in this region should be spoken of as a "hybrid swarm." The specimens demonstrate a much greater affinity with *C. latrans* than with *C. rufus* or any other species, and may be properly referred to as coyotes. The subspecies considered by Jackson (1951:271) to inhabit the most proximal geographic area was *C. latrans frustror*. His description of this subspecies as the "largest coyote" was based in part on the examination of specimens that probably represented introgression from *C. rufus*.

Examination of Post-1960 Material from Coastal Texas

The situation within 100 miles of the Texas Gulf Coast is markedly different from that found farther inland. Red wolves survived in the southeastern corner of Texas, east of the Brazos River, through the 1930's and 1940's (pp. 45-46). Some skulls taken farther south along the coast also seem to resemble those of *C. rufus*. Halloran (1961) reported that five supposed red wolves were collected on the Aransas National Wildlife Refuge in 1956, but the listed weights were much less than those of specimens that he said had been taken in 1939-1940. Davis (1966:112-113) depicted the recent range of *C. rufus* as extending all along the Gulf Coast from the Louisiana border to Baffin Bay, but he observed that the species was "on the verge of extinction." Beezley (1967) wrote that red wolves were present on the coastal prairies from Jefferson to San Patricio counties, and farther inland in Harris, Wharton, Colorado, Lavaca, and Victoria counties.

Once again, my analysis of material is divided on a partly arbitrary basis. The localities, many known only to county, are plotted in figure 31 and listed in appendix A (parts 33-38).

Kenedy County.—One of the most interesting of the coastal specimens was taken in December 1961 near Armstrong, Kenedy County, not far north of the Mexican border. The skin and skeleton were reported as *C. rufus* by Paradiso (1965), but with no further comment. The statistical position of the large skull falls well within the limits of the standard red wolf sample (Fig. 37). And yet the specimen was taken in an area beyond the southern edge of the range assigned by Goldman (1944:487) to *C. r. rufus*. Other material, taken south of Nueces Bay in an earlier period, does not approach the Kenedy County skull in size or other characters, and appears

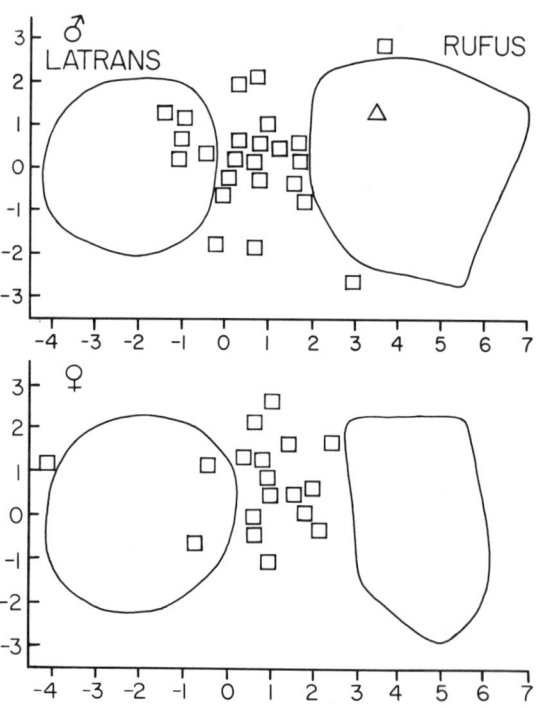

Fig. 37.—Multivariate positions (squares) of individual post-1960 specimens from the central Gulf Coast of Texas, relative to series of *C. rufus* and *C. latrans*. The triangle shows the position of an apparent specimen of *C. rufus* collected in 1961 in Kenedy County, extreme southern Texas.

to represent only *C. latrans* (see Fig. 20). No additional recent specimens from the area are available, but further investigation may yet reveal evidence of the presence of a small, isolated red wolf population.

Central Gulf Coast.—Figure 37 shows the relative positions of skulls of 24(7) males and 18(5) females, taken from 1962 to 1969 in an area including Calhoun, Victoria, Matagorda, Lavaca, Colorado, Austin, and Fort Bend counties. No character clines are apparent in this area, and several counties were the source of both wolflike and coyotelike specimens. Most fall between the extreme limits of *C. rufus* and *C. latrans,* and their statistical distribution is similar to that shown by the former hybrid swarm of central Texas (Fig. 23). Apparently, significant numbers of red wolves survived on the central coast of Texas longer than farther inland. Coyotes have now occupied the same area and hybridized with the wolves, and there is no way of separating the two parent species. Specimens taken in the 1930's and 1940's (Fig. 26) indicate that coyotes were then already beginning to predominate, but evidently red wolf influence persisted. Possibly, small groups of relatively unmodified *C. rufus* still are present in the area, but it would be difficult to pinpoint their locations.

Harris County.—The vicinity of the Addicks Reservoir, just west of Houston, was the

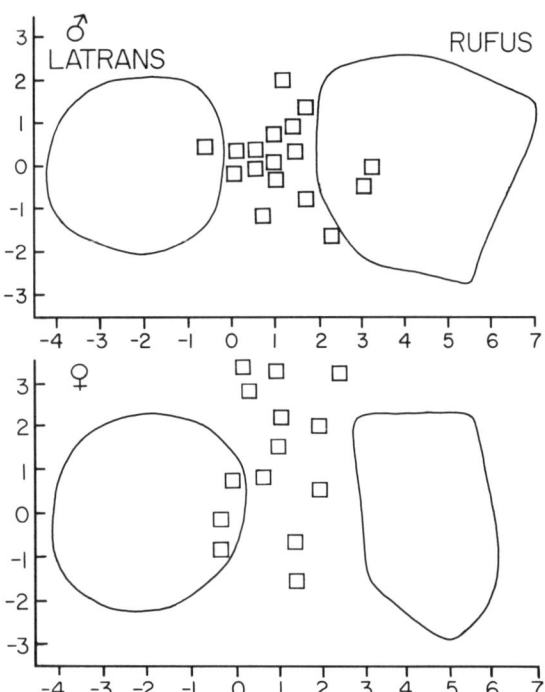

Fig. 39.—Multivariate positions (squares) of individual post-1960 specimens from western Brazoria County, Texas, relative to series of *C. rufus* and *C. latrans.*

source of 12(2) males taken from 1964 to 1971. Seven of these have a smaller D^2 separation from *C. rufus* than from *C. latrans.* The over-all statistical distribution (Fig. 38) suggests that the coastal hybrid population is established in the area, but that red wolf influence still is strong. On the basis of investigation of the ecology, behavior, morphology, and allelic frequency of canids taken mainly in southeastern Harris County, Shaw (1975: 95) suggested that the population there represented hybridization between *C. rufus* and *C. latrans.*

Western Brazoria County.—From 1968 to 1971, 17(13) males and 14(10) females were collected in the vicinity of the Clemens Prison Farm, south of the town of Brazoria and west of the Brazos River. Their statistical distribution (Fig. 39) is indicative of a hybrid population bridging the gap between *C. rufus* and *C. latrans.*

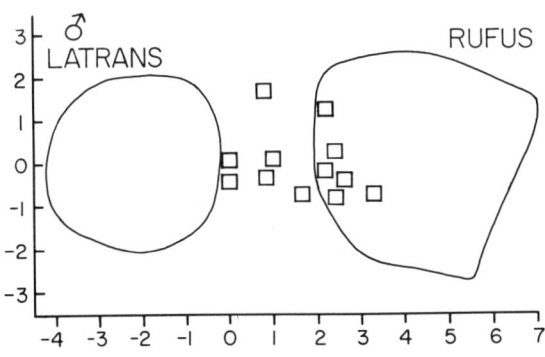

Fig. 38.—Multivariate positions (squares) of individual post-1960 specimens from Harris County, Texas, relative to series of *C. rufus* and *C. latrans.*

Eastern Brazoria County.—Paradiso and Nowak (1972a) reported that typical *C. rufus rufus* still lived in Brazoria County east of the Brazos River. Russell and Shaw (1971b) located red wolves in the same area, through elicitation of howling responses. A total of 11(5) male and 9(4) female adult skulls, some of them not available for Paradiso and Nowak's earlier work, were tested by multivariate analysis. All of this material was obtained between 1966 and 1971 in the east-central part of the county, in the vicinity of Angleton, Liverpool, and Hopkins Mound. Statistical positions (Fig. 40) are concentrated along the lower limits of the standard red wolf sample, not unlike those of older material assigned to the subspecies *C. r. rufus*. Four of the specimens, however, actually have a greater D^2 separation from *C. rufus* than from *C. latrans*. While it thus is evident that gene flow from coyotes has had an effect in this area, the influence of *C. rufus* seems stronger here than farther west on the Gulf Coast. Several other skulls from eastern Brazoria County, not suitable for multivariate analysis, were examined and found to possess typical red wolf characters. One of these skulls was 241.0 millimeters in greatest length. Recent reports from field personnel, however, suggest that red wolf influence in this area has declined, and conservation efforts there have been largely discontinued.

Jasper, Tyler, Liberty, and western Chambers counties.—Red wolves held out for many years in the area of the Big Thicket to the northeast of Houston. Currently, however, it is not likely that the species, in unmodified form, is present in the area. According to recent reports, it seems that the expanding coyote population has occupied the woodlands of southeastern Texas in moderate numbers (Russell and Shaw, 1971b; Glynn Riley, U.S. Fish and Wildlife Service, pers. comm.). The status of wild *Canis* here is not well understood, and I have examined no specimens from within the borders of the Big Thicket itself. The positions of 3(1) males taken in 1970 on the eastern side of the Thicket, two near Fred, Tyler County, and one near New Blox, Jasper County, indicate that the influence of *C. rufus* is still strong in the area (Fig. 41). Skulls of 11 males and 10 females were taken from 1965 to 1971 along the lower edge of the Big Thicket in southern Liberty and western Chambers counties. This area contains a mixture of coastal prairie and woodland habitat. Some of the specimens are indistinguishable from red wolves collected many years ago in the south-central states, but others are coyotelike or intermediate in characters (Fig. 41). Apparently then, coyotes have spread through east Texas, and have come into contact with surviving red wolves on the coastal prairies where hybridization probably has occurred.

Jefferson and eastern Chambers counties.—Even while the disappearance of red wolves over most of their former range was being

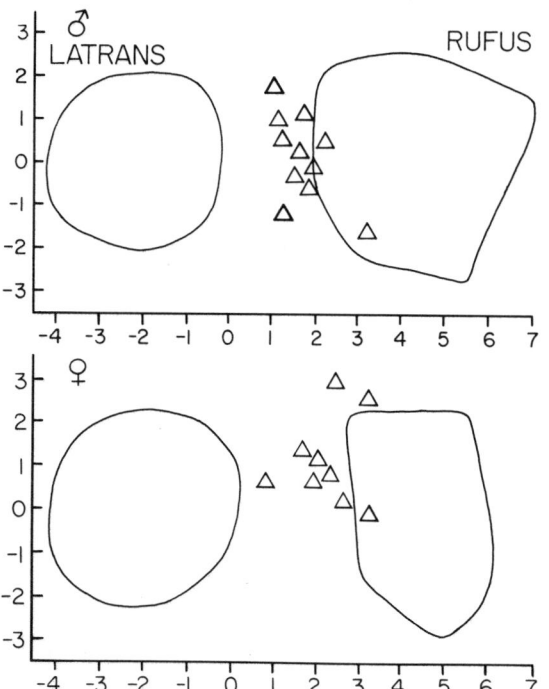

Fig. 40.—Multivariate positions (triangles) of individual specimens taken from 1966 to 1971 in eastern Brazoria County, Texas, relative to series of *C. rufus* and *C. latrans*.

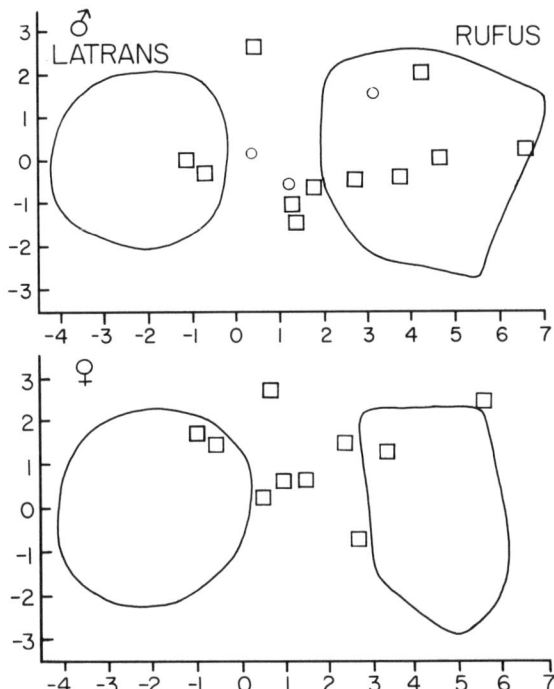

FIG. 41.—Multivariate positions of individual specimens taken from 1965 to 1971 in Jasper and Tyler counties, Texas (open circles), and in southern Liberty and western Chambers counties (squares), relative to series of *C. rufus* and *C. latrans*.

generally acknowledged, there were continuous reports of their survival on the coastal prairies and marshes of extreme southeastern Texas between Galveston Bay and Sabine Lake. Paradiso (1965) reported that seven large skulls, collected in 1963 or early 1964 on the Anahuac National Wildlife Refuge, southeastern Chambers County, were unquestionably those of *C. rufus*. Pimlott and Joslin (1968) wrote that in 1965 they located three packs of red wolves in the area east of Galveston Bay between Highway 73 and the Gulf Coast. Paradiso and Nowak (1972a) considered that specimens collected between 1963 and 1969 demonstrated the survival of the subspecies *C. rufus gregoryi* in the area.

For the purposes of this paper, skulls of 15(3) males and four females taken from 1963 to 1970 were tested by multivariate analysis. All skulls are from animals killed in eastern Chambers and southern Jefferson counties, the area between Galveston Bay and Sabine Lake, and south of the line formed by U.S. Interstate Highway 10 and Texas Highway 73. The relative statistical positions of these specimens (Fig. 42) fall predominantly within the range of variation of standard *C. rufus*, and confirm the presence of the species, in apparently unmodified form, in extreme southeastern Texas (see also measurements in appendix B, part 7).

The suggestion that the red wolf survived in southeastern Texas until about 1970 was supported by field studies carried out in that area. Glynn Riley, an agent of the U.S. Fish and Wildlife Service, who was locally responsible for red wolf management from 1969 to 1973, said (pers. comm.) that the species was present in moderate to high numbers in Chambers and Jefferson counties. In Septem-

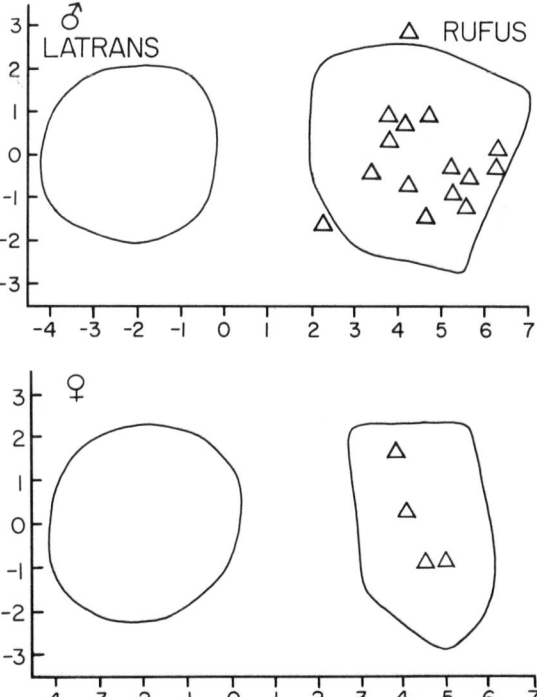

FIG. 42.—Multivariate positions (triangles) of individual specimens taken from 1963 to 1970 in Jefferson and eastern Chambers counties, extreme southeastern Texas, relative to series of *C. rufus* and *C. latrans*.

ber 1972 he estimated the population south of Interstate Highway 10 to consist of about 150 individuals. Russell and Shaw (1971b) wrote that in 1970 they located a dense red wolf population in southern Chambers and Jefferson counties, through elicitation of howling responses. Shaw (1975) reported a relatively high population density of *C. rufus gregoryi* in the area during his field work from June 1971 to September 1972, and found the animals so identified to have gross morphological and behavioral characters distinct from those of *C. latrans*.

Probably an important factor in the survival of these wolves was their geographic location. Far to the south along the Gulf Coast, they were well removed from the original and subsequently expanding coyote population. The temporary existence in eastern Texas of a large zone in which wild *Canis* was rare or absent (see pp. 52-53), helped to prevent genetic exchange between coyotes to the north and the surviving wolves on the coast. To the west, Galveston Bay and the Houston metropolitan area limited the spread of the hybridization process that engulfed most of the coast.

In my dissertation (Nowak, 1973:146) I expressed hope that conservation efforts might help to maintain the unmodified population of *C. rufus* in southeastern Texas. Even before I wrote those words, however, the situation probably had deteriorated beyond the point of practical control. Indeed, there is good evidence that canids with coyotelike characteristics were already in Jefferson and eastern Chambers counties in the 1960's, and that the skulls of such animals may never have been submitted for examination, because persons obtaining specimens thought that only "wolves" were wanted (Carley and McCarley, 1976). This bias probably was not an overriding factor, however, since both wolflike and coyotelike skulls were received from many other parts of Texas (see Figs. 36-41). While it is likely that a portion of the population of wild *Canis* in Jefferson and eastern Chambers counties in the 1960's did not represent unmodified *C. rufus*, I continue to think that this portion was small relative to other areas of coastal Texas.

In any event, the situation does appear to have worsened in the late 1960's and early 1970's. Conservation efforts by the U.S. Fish and Wildlife Service, including some trapping of non-wolflike canids in the vicinity, did not prevent the continued decline of the red wolf. By 1974, if not sooner, field personnel recognized that a substantial number of the wild canids south of Interstate Highway 10 were not wolves (Curtis J. Carley, Project Leader, Red Wolf Recovery Program; pers. comm.). Skulls of 13 males and 12 females, collected from 1973 to 1975 in the same area as those represented in figure 42, were evaluated by multivariate analysis. The statistical positions of these new specimens are shown in figure

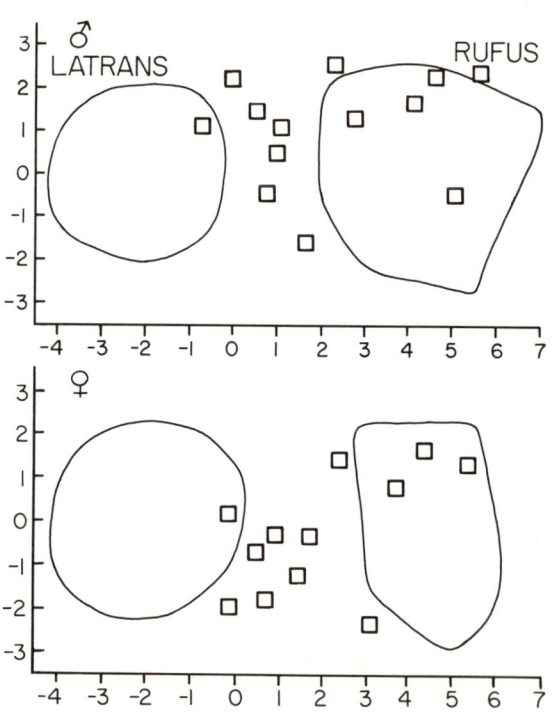

Fig. 43.—Multivariate positions (squares) of individual specimens taken from 1973 to 1975 in Jefferson and eastern Chambers counties, extreme southeastern Texas, relative to series of *C. rufus* and *C. latrans*.

43, and the general pattern resembles that of material taken just to the north from 1965 to 1971 (Fig. 41). Thus it is evident that coyotelike animals have now completely occupied the former range of *C. rufus* in Texas, and that no red wolf populations (though probably some individuals) remain free from significant genetic influence by *C. latrans*.

When it was recognized that the last red wolf population was being swamped through hybridization, the U.S. Fish and Wildlife Service accelerated a program to live-capture some individuals that did appear to represent pure *C. rufus,* and to place these animals in a breeding pool at the Point Defiance Zoo at Tacoma, Washington. In November 1977 this facility contained 29 adults, all wild-caught (including three from Cameron Parish, Louisiana), and 13 young produced during the spring of 1977 (Norman R. Winnick, Director, Red Wolf Captive Breeding Program, Point Defiance Zoo; pers. comm.). On 13 December 1976 a pair from this group had been released on Bulls Island, Cape Romain National Wildlife Refuge, South Carolina. Both animals were recaptured on 22 December 1976 after one had crossed to the mainland, and the female subsequently died of natural causes. Another release was planned for January 1978. Although Bulls Island is too small (5,000 acres) to support a viable pack of wolves, it is hoped that procedures can be developed there that will aid possible future reintroductions in other parts of the original range of *C. rufus*. At present, however, the wild wolf populations that once inhabited the southeastern quarter of North America are not in existence.

SYSTEMATIC DESCRIPTIONS

ORDER CARNIVORA
FAMILY CANIDAE
Genus **Canis** Linnaeus

1758. *Canis* Linnaeus, Systema Naturae, 10th ed., p. 38. Type, *Canis familiaris* Linnaeus.
1816. *Thos* Oken, Oken's Lehrbuch der Naturgeschichte, pt. 3 (Zoologie), sect. 2, p. 1037. Type, *Thos vulgaris* Oken (=*Canis aureus* Linnaeus).
1839. *Lyciscus* Hamilton-Smith, in The naturalist's library (edit. Jardine), 25:160. Type, *Canis latrans* Say.
1918. *Aenocyon* Merriam, Univ. California Publ. Bull. Dept. Geol., 10:532, 20 April. Type, *Canis dirus* Leidy, subsequent designation by Hay, Second bibliography and catalogue of the fossil Vertebrata of North America, 2:501, 27 January 1930.

Geological distribution.—Middle Pliocene to Recent.

Geographical distribution.—In historical time, the species of *Canis* have been distributed as follows:

Canis familiaris Linnaeus, 1758 (including *C. dingo* Blumenbach, 1780, of Australia), the domestic dog, throughout the world, usually in association with man;

Canis lupus Linnaeus, 1758, the gray or timber wolf, in most of North America and Eurasia, and on associated northern islands;

Canis rufus Audubon and Bachman, 1851, the red or southern wolf, in southeastern North America;

Canis latrans Say, 1823, the coyote or prairie wolf, primarily in western North America;

Canis aureus Linnaeus, 1758 (including *C. lupaster* Hemprich and Ehrenberg, 1832, of northern Africa), the golden jackal, from central Africa to India and southeastern Europe;

Canis adustus Sundevall, 1846, the side-striped jackal, in eastern, southern, and western Africa;

Canis mesomelas Schreber, 1778, the black-backed jackal, in eastern and southern Africa; and

Canis simensis Ruppell, 1835, the Abyssinian wolf, in the mountains of Ethiopia.

If most species assigned to the genus *Dusicyon* are actually referable to *Canis*, as suggested by Langguth (1975), the Recent geographical distribution of *Canis* would include much of South America. Pleistocene fossils of *Canis* have been reported from South America, as well as from North America, Eurasia, and Africa.

Relationship with other Recent genera.—The Recent Canidae often have been divided into three subfamilies: Caninae, with the genera *Canis*, *Alopex*, *Fennecus*, *Vulpes*, *Urocyon*, *Dusicyon*, *Chrysocyon*, *Atelocynus*, *Cerdocyon*, and *Nyctereutes*; Simocyoninae, with the genera *Cuon*, *Lycaon*, and *Speothos*; and Otocyoninae, with the genus *Otocyon*. Van Valen (1964) suggested that *Otocyon* does not warrant subfamilial distinction from the Caninae, and most workers seem to be following this viewpoint. Clutton-Brock, Corbett, and Hills (1976) proposed abandoning the use of all subfamily names for the Recent Canidae. These authors also placed *Urocyon* and *Fennecus* within the genus *Vulpes*, and, like Osgood (1934) and Simpson (1945:109), included *Cerdocyon* and *Atelocynus* in the genus *Dusicyon*. Langguth (1975), however, retained the genera *Cerdocyon* and *Atelocynus*, put all but one of the species of *Dusicyon* in the genus *Canis*, and placed the remaining species (*Dusicyon vetulus*) in the genus *Lycalopex*.

Osgood (1934) prepared a systematic key in an attempt to demonstrate the common affinity of all South American Caninae and their collective distinction from *Canis*. The South American genera (including *Urocyon*)

were said to be characterized by a reduced upper carnassial, with a length always less than the combined length of the two upper molars; an occipital shield depressed at the apex; a usually flattened or lyrate sagittal crest (except in *Chrysocyon*); relatively small upper incisors, the inner pair imperfectly or not trifid; and long, slender canine teeth. These characters, however, may be only relative expressions of trends found throughout the Caninae. For example, in some North American coyotes the saggital crest is flattened and lyrate, and P4 is shorter than the combined length of M1 and M2.

Study of canid karyotypes thus far seems to support the composition of the genus *Canis* as set forth above under "geographical distribution." Each species of *Canis* that has been analyzed (*C. familiaris, C. dingo, C. lupus, C. rufus, C. latrans,* and *C. aureus*) has a diploid chromosome number of 78. Species of other genera within the Caninae, including *Vulpes, Fennecus, Alopex, Urocyon, Chrysocyon, Nyctereutes,* and *Otocyon,* have been found to have diploid numbers different from that of *Canis*. The species *Dusicyon vetulus* (placed in the genus *Lycalopex* by Langguth, 1975) has a diploid number of 76 (Wurster and Benirschke, 1968; Chiarelli, 1975).

The production of hybrids provides further evidence of the validity of the presently accepted structure of the genus *Canis*. Viable hybrids have been reported between *C. familiaris* and *C. lupus, C. familiaris* and *C. latrans, C. familiaris* and *C. aureus, C. familiaris* and *C. dingo, C. latrans* and *C. aureus, C. latrans* and *C. lupus, C. lupus* and *C. dingo* (Gray, 1972:45-51), and *C. latrans* and *C. rufus* (this paper). There are few data on interbreeding between members of *Canis* and other recognized genera, and no evidence that viable hybrids ever have been produced (see Chiarelli, 1975:50).

Remarks.—Determination of the actual origin of the genus *Canis* is not within the scope of this paper. One widely held view is that the genus *Tomarctus* of the Miocene and early Pliocene is near the ancestral line of *Canis* and other modern Caninae. Simpson (1945:109, 222) placed *Tomarctus* within the Caninae, but noted that it might represent the ancestry of another subfamily. Matthew (1930:131) and Vanderhoof and Gregory (1940:145) considered *Tomarctus* to occupy a central position from which both the various borophagine dogs and the modern canines arose. Green (1948) concluded that certain species of *Tomarctus* were in the direct line of the borophagines, whereas other species probably gave rise to *Canis* and possibly to *Vulpes*. Williams (1962) referred *Tomarctus* to the Borophaginae and placed all but one of its named species in the genus *Aelurodon*. He suggested that some lesser known canine of the late Miocene or early Pliocene, such as *Leptocyon vafer* (Leidy), was the ancestor of the Quaternary Caninae. Presumably then, according to this interpretation, the lineage leading to the living Caninae has been distinct from other subfamilial groupings at least since the late Miocene, and *Canis* has descended from small, foxlike ancestors. At present, the number of reported Pliocene canines is small, and the particular point at which *Canis* arose has not been established.

It is true that many North American specimens from the Pliocene and even earlier times, some of them representing animals larger than a modern wolf, were named as species of *Canis*. Nearly all of these names, however, now have been referred to various other genera, and they seem to have no bearing on the lineage of *Canis* (Hay, 1902:769-776, and 1929-1930:488-512 may be consulted for these named kinds and their places of description). Most recently, the species *Canis texanus* Troxell, 1915 was referred to the genus *Protocyon* by J. L. Kraglievich (1952: 621).

Only three Pliocene species of *Canis* are now known from North America. These include *C. davisii* from southwestern Nevada (Merriam, 1911:242-243) and southeastern

Oregon (Shotwell, 1970:70), and *C. condoni* from northeastern Oregon (Shotwell, 1956: 733). Shotwell (1970) noted that additional material might prove these two species to be synonymous. *Canis lepophagus,* discussed in detail below, is known from a number of late Pliocene and early Pleistocene sites.

Of the various specific names applied to North American Pleistocene fossils of *Canis,* the following 17 have not hitherto been placed in other genera or formally reduced to synonyms or subspecies:

> *C. cedazoensis* Mooser and Dalquest, 1975*
> *C. lepophagus* Johnston, 1938*
> *C. latrans* Say, 1823*
> *C. andersoni* Merriam, 1910
> *C. riviveronis* Hay, 1917
> *C. caneloensis* Skinner, 1942
> *C. irvingtonensis* Savage, 1951
> *C. edwardii* Gazin, 1942*
> *C. rufus* Audubon and Bachman, 1851*
> *C. priscolatrans* Cope, 1899
> *C. armbrusteri* Gidley, 1913*
> *C. lupus* Linnaeus, 1758*
> *C. milleri* Merriam, 1912
> *C. familiaris* Linnaeus, 1758*
> *C. petrolei* Stock, 1938
> *C. dirus* Leidy, 1858*
> *C. ayersi* Sellards, 1916.

The status of all of these names is discussed in the following pages. Each of those marked above with an asterisk (*) is maintained as a specific name in this paper.

Canis cedazoensis Mooser and Dalquest

1975. *Canis cedazoensis* Mooser and Dalquest, Jour. Mamm., 56:787.

Holotype.—Right maxillary fragment containing P3 to M1 and the alveolus of M2; no. 9780, Midwestern State University Department of Biology; Arroyo Cedazo, 3 kilometers SE City of Aguascalientes, Aguascalientes, Mexico.

Geological distribution.—Early Rancholabrean.

Geographical distribution.—Known only from the type locality.

Description.—A small canid, larger than any North American fox, but smaller than *C. latrans;* M1 relatively small, with pronounced buccal cingulum (for more detail, see Mooser and Dalquest, 1975:787-788).

Comparison with C. latrans.—Smaller; M1 with relatively smaller medial section.

Remarks.—The single available specimen is too fragmentary, and the teeth too heavily worn, to allow the same kind of account given to other species in this paper. In general I am in accord with the description and comparisons provided by Mooser and Dalquest (1975:787-788), and with their conclusion that the specimen does not represent *C. latrans.*

Record of occurrences.—Type locality; early Rancholabrean (probably Illinoian); specimen examined: holotype, Midwestern State University Department of Biology 9780 (formerly no. FC 634 in collection of O. Mooser); measurement in appendix B (part 8).

Evolutionary position.—On the basis of the one available specimen, the relationships of *C. cedazoensis* can not be carefully assessed. The species might represent an aberrant line that separated from *C. latrans* in the Irvingtonian, or it could be a surviving element of one of the smaller-sized populations of the Blancan *C. lepophagus.*

Canis lepophagus Johnston

1938. *Canis lepophagus* Johnston, Amer. Jour. Sci., ser. 5, 35:383.

Holotype.—Skull without mandibles; no. W. T. 881, Panhandle Plains Museum; stratum no. 2, Harold Ranch, North Cita Canyon, center of west half of sec. 164, block 6, Randall County, Texas.

Geological distribution.—Blancan.

Geographical distribution.—Known from California, Florida, Idaho, Kansas, Nebraska, and Texas.

Description.—A small coyotelike canid;

skull small with mostly narrow proportions; rostrum elongated and narrow; braincase relatively small and little inflated dorsoposteriorly; frontal and supraoccipital shields relatively broad; sagittal crest prominent; mandible long and narrow; premolars and molars with trenchant, laterally compressed cusps; P4 with prominent deuterocone; M1 with pronounced buccal cingulum; p4 with prominent second cusp.

Comparison with C. latrans.—Skull averaging smaller, but overlapping in most measured dimensions; braincase relatively smaller and less inflated dorsoposteriorly; frontal and supraoccipital shields relatively broader; sagittal crest more prominent; some mandibles relatively deeper; dentition usually more crowded and with more trenchant, laterally compressed cusps; metaconule on M1 less separated from protocone; p4 usually with more prominent second cusp, and with third cusp and posteromedial cingulum more reduced (if they are present); anterior margin of m1 usually more nearly vertical. Bjork (1970:14) and Kurten (1974) discussed a number of postcranial differences between the two species.

Other comparison.—See account of *C. edwardii*.

Remarks.—Johnston (1938) wrote that *C. lepophagus* differed from *C. latrans* in having more prominent sagittal and lambdoidal crests, less expanded braincase, deeper mandible, and lesser distance between premolars. Subsequently, other workers assigned additional specimens to *C. lepophagus*, but not all of this material shared the characters cited by Johnston. Hibbard (1938:243) wrote that a specimen from the Rexroad fauna, which he later (1941a:268-269) referred to *C. lepophagus*, represented a "small dog with a light tapering ramus." Hibbard (1941a) stated that *C. lepophagus* from this fauna was characterized by small size and relatively narrow teeth. Fine (1964) wrote that a mandibular fragment, assigned later by Bjork (1970:14) to *C. lepophagus*, was "from a lightly built coyote having a jaw quite long for its depth as compared with those of Recent specimens." Bjork (1970) described *C. lepophagus* from the Hagerman local fauna as having a long and slender lower jaw, and dentition more slender than that of *C. latrans*. Giles (1960) subjected six mandibles of *C. lepophagus* from Cita Canyon to a multivariate analysis involving five measurements. The results suggested to him that *C. lepophagus* was subspecifically, but not specifically, distinct from Recent *C. latrans*.

According to Kurten (1974:27-28), "*Canis lepophagus* of the Blancan differs from the Recent and Rancholabrean *C. latrans* in its shorter distal limb segments, more tapering snout, relatively larger M2 (this character is however retained in some Recent populations), narrower p2-p4, and shorter m1 (relative to its own width, and also to the length of p4)." He considered all coyotelike material from the Blancan, except from the latest part of this age, to represent *C. lepophagus*, and that this species formed a stage in the evolutionary line leading to *C. latrans*. Material from most of the localities listed below under "record of occurrences," was referred by Kurten to *C. lepophagus*.

I recognize most material previously assigned by others to *C. lepophagus* as representing a single variable species, separate from and ancestral to *C. latrans*. Hibbard (1941b) thought that *C. lepophagus* had as large a geographic range as Recent *C. latrans*. Considering this likelihood, plus the long period of time through which material of *C. lepophagus* is distributed, considerable variation would be expected. The three skulls from Cita Canyon reported by Johnston (1938), plus one other obtained at the same site, are the only specimens of *C. lepophagus* upon which meaningful measurements can be made of elements other than the teeth and lower jaws. And, as listed above, there are differences between these skulls and those of *C. latrans*. The mandibles of *C. lepophagus* from Cita Canyon overlap in size and proportion those of Recent *C. latrans* (Fig. 44), but average deeper. Most specimens from other sites are mandibles, some of them relatively shallow. Differentiating characters of the lower dentition, as listed in the above "comparison with *C. latrans*," are present in most of the Cita Canyon mandibles and in most of the other specimens that can be evaluated. These characters are matched in some individuals of *C. latrans*, but their presence in specimens of *C. lepophagus* from several major sites suggests affinity among the populations represented.

The type locality of *C. lepophagus*, Cita Canyon, now is recognized to contain an Aftonian (Hibbard, 1970:414) or late Blancan (Kurten, 1974:5) fauna, and thus is among the youngest sites from which the species has been reported. Since no specimens from contemporary sites are complete enough for definite assignment to *C. latrans*, but since some do have characters typical of *C. lepophagus*, all coyotelike material from the Blancan is referred below to *C. lepophagus*.

Record of occurrences.—The following list is arranged alphabetically by state, and geographically (north to south, west to east) within states. Specimens examined by me are identified by element, museum number, or both; and selected measurements are found in appendix B (part 9) and appendix C (part 1). Occurrences also are shown on the map in figure 45.

CALIFORNIA.—Sacramento Valley, Sacramento River, Tehama County; late Pliocene; as *Canis* (Vanderhoof, 1933:383); m1, UCMP 29828.

FLORIDA.—Santa Fe River IB, Gilchrist County; Blancan (Nebraskan); as *C.* cf. *lepophagus* (Webb, 1974b:17); five mandibular fragments, UF 10423, 10424, 10836, 10837, 10858. Two of these specimens are comparatively large and deep, and suggestive of the Cita Canyon material.

IDAHO.—Grand View, Owyhee County; late Blancan (Hibbard, et al., 1965); as "*Canis* n. sp." (Schultz, 1938a:297); two mandibular fragments, LACM 118-1246, 1343. The specimens are comparatively large and deep; the p4 has a prominent second cusp, but no third cusp or posteromedial cingulum.

Hagerman local fauna, Twin Falls County; early Blancan (late Pliocene); as *C. lepophagus* (Bjork, 1970:13-16); four maxillary fragments, USNM 25136, UMMP V52280, V54995, V56401; seven mandibular fragments, USNM 25131, UMMP V45222, V50249, V53519, V53817, V53910, V56282; two M1, UMMP V50000, V56034; M1 fragment, UMMP V57016; p2, UMMP V56809; two p3, UMMP V50008, V51052; p4 fragment, UMMP V53817; m1,

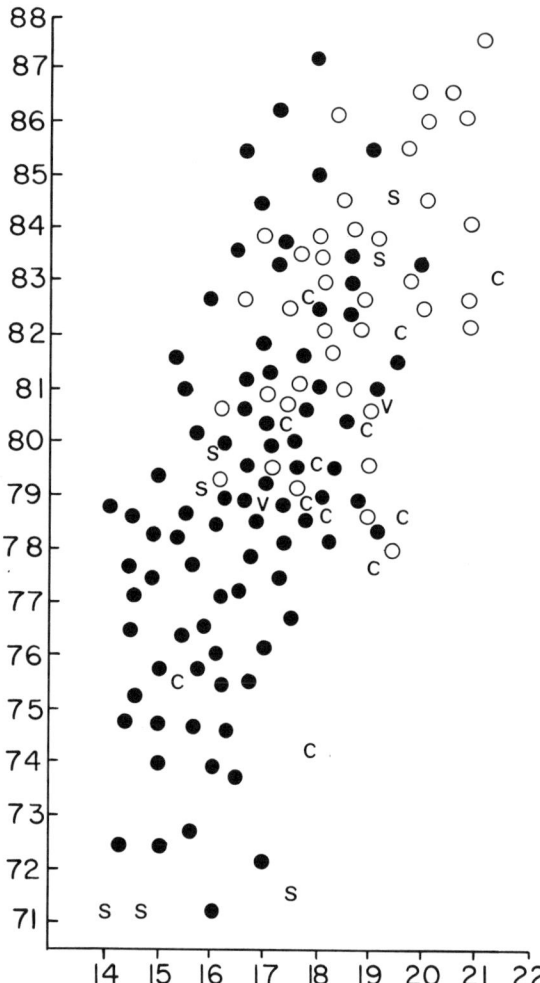

Fig. 44.—Scatter diagram comparing length and depth of mandible in certain specimens of *C. lepophagus* and *C. latrans*. The vertical axis indicates measurement of the distance from the anterior edge of the alveolus of p1 to the posterior edge of the alveolus of m3. The horizontal axis indicates measurement of the minimum depth from the dorsal surface of the mandible between p3 and p4 to the ventral surface of the mandible. Black dots, Recent *C. latrans lestes;* open circles, Pleistocene *C. latrans orcutti;* c, *C. lepophagus* from Cita Canyon; s, *C. lepophagus* from other sites; v, *C. latrans irvingtonensis.*

Fig. 45.—Map showing localities of *C. lepophagus* (black dots) and *C. armbrusteri* (triangles). Because of the scale of the map, it was not possible to plot all localities in crowded areas.

USNM 25132; m1 fragment, UMMP V52757; postcranial elements, UMMP. As explained by Bjork (1970), the Hagerman material indicates an animal smaller than the average Recent *C. latrans*.

KANSAS.—Rexroad fauna, Meade County State Park, Meade County; early Blancan (late Pliocene);

as *C. lepophagus* and *Canis* sp. (Hibbard, 1941a: 268-269); three mandibular fragments, KU 3914, 4602, 4603; M1, m2, KU 3915; P4, M1, UMMP 37132. The mandibles are relatively small and shallow, and most of the teeth are small with trenchant cusps. Hibbard wrote that large size distinguished an M1 and m2, found at the site, from *C. lepophagus*. Although these teeth are larger than the others in the Rexroad fauna, they are within the size range of *C. lepophagus* from other sites.

Rexroad fauna, Keefe Canyon, Meade County; early Blancan (late Pliocene); as *C. lepophagus* (Hibbard and Riggs, 1949:838); m1, KU 7692.

Deer Park local fauna, Meade County; late Blancan (Aftonian); as *Canis* sp. (Hibbard, 1956: 172); two m1, UMMP 31945.

NEBRASKA.—Sand Draw, 6 mi. N Ainsworth, Brown County; late Blancan (Aftonian: Hibbard, et al., 1965); as *C.* cf. *latrans* (McGrew, 1944:53), as *C. lepophagus* (Skinner and Hibbard, 1972:107).

Broadwater quarry, north side of North Platte River, Morrill County; late Blancan (Aftonian); as "*Canis* sp. (near *C. latrans* Say)" (Barbour and Schultz, 1937:4), as "*Canis* sp. (Coyote)" (Schultz and Stout, 1948:563); two skulls, UN 26111, 26112; mandibular fragment, UN 26116; P4, M1, M2, UN 26113. Assignment herein to *C. lepophagus* is uncertain on the basis of morphology alone, as the cranial parts are in too poor a condition for evaluation of specific characters. A measurement of depth on the mandible is not possible, but it is relatively thick lateromedially, and has the massive appearance attributed to *C. lepophagus* by Johnston (1938). The teeth are approximately the same size and have the same structure as those of the Cita Canyon specimens.

Lisco quarry, north side of North Platte River, Garden County; late Blancan (Aftonian); as "*Canis* sp. (Coyote)" (Schultz and Stout, 1948:563); skull, UN 26107; mandible, UN 26114. This material offers some evidence for recognizing that skulls of the Cita Canyon kind, as well as small mandibles from other sites, represent the same species. The Lisco mandible is comparatively small and shallow, with trenchant, laterally compressed cusps on the teeth, and with a high second cusp on p4. The skull also is small, but has a prominent sagittal crest, as do the Cita Canyon skulls.

TEXAS.—Red Coral, Proctor Ranch, Oldham County; late Blancan; "within the variation range of *C. lepophagus*" (Kurten, 1974:5).

Cita Canyon, Randall County; late Blancan (Aftonian: Hibbard, 1970:414); as *C. lepophagus* (Johnston, 1938); four skulls, maxillary fragment, 32 mandibles and fragments thereof, various isolated teeth and postcranial elements, PPM.

Blanco local fauna, near Mount Blanco, Crosby County; Blancan (latest Pliocene or earliest Pleistocene); as *C. lepophagus* (Dalquest, 1975:22, 47).

Beck Ranch, Scurry County; early Blancan; as *C. lepophagus* (Kurten, 1974:5).

Red Light Bolson, southeastern Hudspeth County; Blancan (Nebraskan); as *C.* cf. *lepophagus* (Akersten, 1970:86; 1972:19-24).

Evolutionary position.—The small size and the dental characters of some specimens of *C. lepophagus* suggest affinity to the Old World jackals and to the foxes (*Vulpes*). The other two recognized species of Pliocene *Canis*, *C. davisii* and *C. condoni*, also were described as being foxlike in size and certain other details. Possibly then, the lineage of *Canis* separated from that of *Vulpes* in the early or middle Pliocene. Older specimens of *C. lepophagus* are small and especially indicative of descent from foxlike ancestors. The species *C. latrans* probably evolved from within the complex assigned above to *C. lepophagus*. Perhaps the shallow-jawed material represents the ancestral element of *C. latrans*, whereas the more massive Cita Canyon specimens represent another line.

In my dissertation (Nowak, 1973:164), I agreed with Johnston (1938) in recognizing *C. lepophagus* as an ancestral coyote, but unlike him I did not consider it to be in the direct phylogenetic line of the modern wolves. My opinion was based in part on acceptance of a late Blancan age for the type of *C. edwardii* and several other early specimens of the wolf line. Since the most wolflike specimens of *C. lepophagus*, especially those from Cita Canyon, also were late Blancan in age, but were morphologically distinct from the supposedly contemporary wolf material, they could hardly have been ancestral to *C. edwardii*. New information (Johnson, Opdyke, and Lindsay, 1975; Kurten, 1974) suggests that the type of *C. edwardii*, as well as most other early North American specimens that apparently represent the wolf line, actually are Irvingtonian in age. The one exception is a Blancan specimen from Miñaca Mesa, Chihuahua, which consists of a mandible about the size of that of a small wolf, but which has certain dental characters resembling those of some specimens of *C. lepophagus* (see account of *C. edwardii* below). There thus is a possibility that this mandible represents a transitional phase through which the wolf line did evolve from a population of *C. lepophagus*.

Canis latrans Say

1823. *Canis latrans* Say, in Long, Account of an expedition from Pittsburgh to the Rocky Mountains performed in the years 1819 and 1820, 1:168. Type locality, Engineer Cantonment, about 12 mi. SE present town of Blair, Washington County, Nebraska, on west bank of Missouri River.

1910. *Canis orcutti* Merriam, Univ. California Publ. Bull. Dept. Geol., 5:391. Type from Rancho La Brea, Los Angeles County, California. Valid as a subspecies of *C. latrans*.

1910. *Canis andersoni* Merriam, Univ. California Publ. Bull. Dept. Geol., 5:393. Type from Rancho La Brea, Los Angeles County, California. A synonym of *C. latrans orcutti*.

1917. *Canis riviveronis* Hay, Florida State Geol. Surv. Rept., 9:59. Type from Vero, Indian River County, Florida. Valid as a subspecies of *C. latrans*.

1942. *Canis caneloensis* Skinner, Bull. Amer. Mus. Nat. Hist., 80:163. Type from Papago Springs Cave, Santa Cruz County, Arizona. Valid as a subspecies of *C. latrans*.

1951. *Canis irvingtonensis* Savage, Univ. California Publ. Bull. Dept. Geol. Sci., 28:231. Type from Irvington, Alameda County, California. Valid as a subspecies of *C. latrans*.

1961. *Canis latrans harriscrooki* Slaughter, Jour. Mamm., 42:505. Type from Lewisville site, Denton County, Texas. Valid as a subspecies of *C. latrans*.

In addition to those listed above, 22 names based on Recent specimens are available for use at the subspecific level. These are to be found in the systematic revision by Jackson (1951), and also are listed by Hall and Kelson (1959:843-846).

Type.—None designated.

Geological distribution.—Irvingtonian to Recent.

Geographical distribution.—Pleistocene and early Recent records from Alberta, Alaska, Arizona, Arkansas, California, Colorado, Florida, Idaho, Illinois, Indiana, Iowa, Kansas, Maryland, Mississippi, Missouri, Nebraska, Nevada, New Mexico, Oklahoma, Oregon, Pennsylvania, Texas, Utah, Wisconsin, Wyoming, Aguascalientes, Estado de Mexico, Nuevo Leon, Oaxaca, and Puebla; original range in historical time included most of western half of North America and the plains region as far east as southern Wisconsin, northwestern Indiana, western Arkansas, and central Texas; see also "remarks" below.

Description.—Size small to medium for the genus; skull usually small with mostly narrow proportions; rostrum elongated, narrow, and shallow; braincase relatively large and well inflated dorsoposteriorly, often broader at level of parietotemporal sutures than at base; postorbital constriction broad lateromedially and short anteroposteriorly; zygomata usually slender and not widely spreading; orbits large; frontals usually only slightly elevated above rostrum, not prominently convex, and forming relatively narrow and flat shield; temporal ridges usually not sharp, seldom obscuring frontal suture, and sometimes forming lyrate pattern over braincase; sagittal crest seldom prominent, often thickened lateromedially, sometimes flattened; supraoccipital shield comparatively small; external side of occipital usually with thin-walled projection just above foramen magnum; tympanic bullae well inflated; mandible long, narrow, and shallow, ventral edge often rounded; incisors relatively small, upper canines prominent, thin anteroposteriorly, their alveoli set low in premaxillae, their ventral tips normally extending below level of anterior mental foramina when jaws are closed; premolars with trenchant, laterally compressed cusps; P4 with prominent deuterocone and lingual cingulum; M1 with relatively large, deeply sculptured medial section, prominent metaconule well separated from protocone, usually with pronounced buccal and anterior cingula; M2 relatively large, cusps well developed; p2 usually lacking posterior cusp; p3 sometimes with second and third cusp; p4 with second cusp, and usually with pronounced third cusp and posteromedial cingulum extending behind third cusp; m2 and talonid of m1 relatively large, with trenchant cusps. For details on pelage and postcranial skeleton, see Jackson (1951); Grinnell, Dixon, and Linsdale (1937); Hildebrand (1952a, 1952b, 1954); and Merriam (1912).

Comparisons.—See accounts of *C. cedazoensis, C. lepophagus, C. rufus, C. lupus,* and *C. familiaris.*

Remarks.—When the white man first arrived in North America, *C. latrans* had a wide distribution, primarily in the western half of the continent. The exact southern, northern, and eastern limits of its former range are not known. Young (1951:29) thought that the species was originally found only as far as central Mexico, and that movement farther to the south occurred after the introduction of livestock in the region. But Jackson (1951) recognized the presence of three separate subspecies in Central America, and it is questionable whether all of these could have come into existence in only 400 years. Also, according to Young (1951), the coyote did not become established in northwestern Canada and Alaska until the nineteenth and twentieth centuries. But Jackson (1951:265) suspected that the subspecies *C. latrans incolatus* was a very long time resident of Alaska.

Information provided in the first main part of this paper indicated that the original range of the coyote extended at least as far east as southern Wisconsin, northwestern Indiana, western Arkansas, and central Texas. Skeletal remains, identified as *C. latrans,* have been reported from archeological excavations in Crawford County, Illinois (Parmalee and Stephens, 1972); Will County, Illinois (Parmalee, 1962b); Madison County, Illinois (Parmalee, 1959a); Phelps County, Missouri (Parmalee, 1965); and Washington County, Arkansas (Morrison, 1970). Indian sites from farther east have apparently not yielded specimens of *C. latrans,* but in the late Pleistocene the species ranged as far east as Pennsylvania and Florida. Man's extermination of the larger wolves, and disruption of the environment, contributed to an expansion of the coyote's range since the mid-nineteenth century. The species is now established in Ontario, southern Quebec, New England, New York, the Ohio Valley, and the lower Mississippi Valley; and introduced individuals have been reported from all states of the southeast.

Specimens, especially skulls, of *C. latrans* can almost always be easily separated from those of wolves. Hybridization under completely natural conditions seems to have occurred rarely, if ever. The three specimens described on page 11 are the only ones known from the western half of the continent that appear to represent hybridization between *C. lupus* and *C. latrans*. More extensive interbreeding, probably caused by man's influence, has taken place in southeastern Canada. This process has permitted the introgression of wolf genes into the expanding coyote population of the region, and has resulted in a modification of that population (see pp. 21-23). Lawrence and Bossert (1969) referred to the existing coyotes of the northeastern United States as "*C. latrans* var." Specimens taken there in recent years are variable, but, in comparison with coyotes from farther west, are often characterized by larger size, more massive skull, broader and more elevated frontal shield, more prominent sagittal and lamboidal crests, and teeth with less trenchant cusps.

A related process in the south-central states has allowed introgression of genes from *C. rufus* into the coyote population that occupied the region in the twentieth century. Individuals of this population generally closely resemble coyotes from farther north and west in proportions and dentition, but are characterized by larger skull, more prominent sagittal and lambdoidal crests, and more elongate postorbital constriction.

Kurten (1974) grouped all coyotelike material from the Irvingtonian and latest Blancan of North America under the name *Canis priscolatrans*, which he considered to represent an intermediate stage in the evolutionary transition from *C. lepophagus* to *C. latrans*. He acknowledged that there were difficulties in drawing specific boundaries in this sequence, but thought that differences in limb proportions, between Rancholabrean *C. latrans* and available Irvingtonian specimens, supported a separation of species at that point. In addition, *C. priscolatrans* was reported to be considerably larger, on the average, than *C. latrans* or *C. lepophagus*, and to differ from those species in certain dental proportions.

It is important to note that Kurten's (1974) delineation of *C. priscolatrans* was based in part on his inclusion within that species of several large specimens which had been assigned by me (Nowak, 1973) or others to the wolf line. Among this material is the type of *C. priscolatrans*, which I considered not more than subspecifically distinct from *C. rufus*; the type of *C. edwardii*, which Gazin (1942) and I considered to be closely related to *C. rufus*; and specimens from the Inglis IA fauna in Florida which were referred to *C. rufus* by Webb (1974b:17). I continue to think that this material represents the wolf line, and that several other (but certainly not all) Irvingtonian or late Blancan specimens, assigned by Kurten to the coyote line, probably are referable to *C. rufus* or *C. edwardii*. If such specimens are removed from consideration, we find that the Irvingtonian coyotes are not notably larger than those of the Rancholabrean. Although I did not make a detailed analysis of limb bones, as did Kurten, he apparently evaluated only three specimens designated by him as *C. priscolatrans*, one of which was part of the Inglis material referred to *C. rufus* by Webb (1974b:17), and another of which was from a partial skeleton, recovered at Arkalon gravel pit in Kansas, which did not include cranial portions. I do not consider that this sample, or the analysis based thereon, can substantially support a specific division of the coyote line at the Irvingtonian-Rancholabrean boundary, and I continue to think that this line was represented in the Irvingtonian only by the species *C. latrans*.

Kurten (1974:12, 27) also stated that the species *Canis arnensis* of the Villafranchian of Europe closely resembled the North American material he referred to *C. priscolatrans*, but that the former averaged smaller. He observed that future studies might demonstrate that these two species formed part of a single Holarctic coyote population, in which case *C. arnensis* would have to be regarded as a synonym or subspecies of *C. priscolatrans*. I would not agree with the use of the name *C. priscolatrans* for such a population, but since I have not examined any specimens of *C. arnensis* I can not comment on the question of their relationship to North American *Canis*.

Fossil record.—The following list is arranged alphabetically by state and province, and geographically (north to south, west to east) within states, except that Alberta is listed first and Mexican states last. Specimens examined by me are identified by element, museum number, or both; and selected measurements are found in appendix B (part 10) and appendix C (part 2). Occurrences also are shown on the map in figure 46.

ALBERTA.—Medicine Hat; Kansan, Sangamon, Wisconsin; as *C.* cf. *latrans* (Churcher, 1969b:180; Kurten, 1974:7, 10).

ALASKA.—Cripple Creek Mine, near Fairbanks; Wisconsin; as *C. latrans* (Guthrie, 1968:352).

ARIZONA.—Anita, Coconino County; early Irvingtonian (Richard H. Tedford, American Museum of Natural History, pers. comm.); as *C. latrans* (Hay, 1921:633); C1, USNM.

Ventana Cave, Papago Indian Reservation, Pima County; Wisconsin (Hibbard, 1958); as *C. latrans* (Colbert, 1950:132).

Papago Springs Cave, southeast of Sonoita, Santa Cruz County; Wisconsin (Hibbard, 1958); as *C. caneloensis* (Skinner, 1942:163), as *C. latrans cane-*

Fig. 46.—Map showing localities (black dots) of fossil *C. latrans*. Because of the scale of the map, it was not possible to plot all localities in crowded areas.

loensis (Slaughter, 1966b:480), as *C. latrans* (Anderson, 1968:22); skull without mandibles, AMNH 42800. Skinner originally distinguished *caneloensis* as a species, on the basis of its following characters, as compared to the living *C. estor* (=*C. latrans mearnsi*): proportionally wider face, larger bullae, wider P4, less prominent hypocone on M1, and wider M2 with less developed hypocone and protoconule. Now, however, *C. estor* and all other named kinds of Recent coyotes have been arranged as subspecies of *C. latrans* (Jackson, 1951), and it seems unlikely that these differences presently would be considered more than subspecific. Slaughter (1966b: 480) demonstrated that the relative facial width of *caneloensis* fell within the range of variation of Recent *C. latrans*. The additional characters listed by Skinner are matched in large series of Recent coyotes. The interruption of the ridge extending anteriorly from the hypocone of M1 is unusual, but does occur rarely in modern coyotes. In the structure of the occipital, braincase, frontals, and sagittal crest, and in all other features that can be evaluated, the skull is not different from typical western *C. latrans*. I thus concur with Slaughter in reducing *caneloensis* to subspecific rank.

Murray Springs, 1 mi. W Lewis Spring on San Pedro River, Cochise County; late Pleistocene; mandibular fragment, UAriz 2406.

Double Adobe, Cochise County; late Wisconsin; as *C. latrans* (Kurten, 1974:8).

ARKANSAS.—Eddy Bluff shelter, near Springdale, Washington County; early Recent; as *C. latrans* (Morrison, 1970).

Conard fissure, 15 mi. S Harrison, Newton County; Illinoian (Kurten, 1963:100); as *C. latrans* (James H. Quinn, Department of Geology, University of Arkansas, pers. comm.).

Peccary Cave, eastern Newton County; early Recent; as "coyote" (Quinn, 1972:93).

CALIFORNIA.—Samwel Cave, Shasta Lake, Shasta County; Wisconsin (Hibbard, 1958); as *C. latrans* (Graham, 1959).

Hawver Cave, 5 mi. E Auburn, El Dorado County; Wisconsin (Hibbard, 1958); as *C. ochropus* (Stock, 1918:479), as *C. latrans* (Anderson, 1968: 22); cranial fragment, UCMP 11041.

Teichart gravel pit, Sacramento County; late Pleistocene; mandibular fragment, UCMP 85379.

Murphys, Calaveras County; Pleistocene; as *C. latrans* (Whitney, 1879:246), as "*C. latrans*?" (Merriam, 1903:290).

Irvington, Alameda County; Irvingtonian; as *C. irvingtonensis* (Savage, 1951:231); two mandibular fragments, UCMP 38748, 38805; cranial fragment, UCMP 56090; radius, UCMP 38804. According to Savage, *irvingtonensis* has a relatively deeper horizontal ramus, and relatively wider and more closely spaced premolars than *C. latrans*. He compared the two mandibles to 79 specimens of Recent western coyotes, and to 30 jaws of Pleistocene *C. latrans* from Rancho La Brea. He listed eleven dental measurements of *irvingtonensis*, but none on the comparative material, and also none to demonstrate the proportional depth of the Irvington jaws. Although I agree that these mandibles have a more massive appearance than those of most other coyotes, I find all measurements to fall within the range of variation of Recent and Pleistocene *C. latrans* from western North America. The relative depth of the horizontal ramus and width of the teeth also overlap the corresponding dimensions in *C. latrans* (Figs. 44, 47). The premolars are unusually close together, but their spacing varies in *Canis*. In development of the posterior cusps and cingulum on p4, and in other features that can be evaluated,

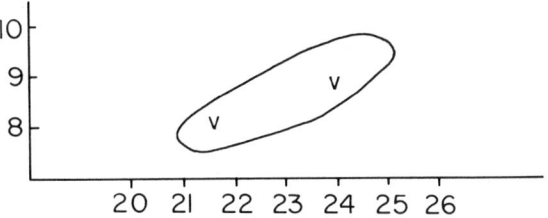

Fig. 47.—Scatter diagram comparing crown length of m1 (horizontal axis) and crown width of m1 (vertical axis), in 40 specimens of Pleistocene *C. latrans orcutti* from Rancho La Brea (range of variation indicated by solid line) and two specimens of *C. latrans irvingtonensis* (v).

the Irvington material matches series of *C. latrans*. Consequently I do not consider this material to represent a separate species. Savage's name may be tentatively maintained as a subspecific designation, *C. latrans irvingtonensis*, until sufficient material is available to more fully evaluate the situation.

Tranquility, Fresno County; late Wisconsin; as *C. latrans* (Kurten, 1974:8).

McKittrick tar seeps, Kern County; Wisconsin; as *C. latrans orcutti* (Schultz, 1938b:165; Giles, 1960:385); 27 skulls and cranial fragments, 16 mandibles and fragments thereof, LACM. Although he recognized considerable variation in cranial and dental characters, Schultz referred all McKittrick coyotes to the subspecies *orcutti*. He observed that some of the skulls were characterized by comparatively more massive dentition, larger size, and a broader muzzle than is to be seen in the living California subspecies, *C. latrans ochropus*.

Maricopa Brea, near Maricopa, Kern County; Wisconsin; 16 skulls and cranial fragments, 23 mandibles and fragments thereof, various isolated teeth, LACM. This hitherto unreported collection closely resembles that of *C. latrans orcutti* from Rancho La Brea and McKittrick.

Carpinteria asphalt, Santa Barbara County; Wisconsin (Hibbard, 1958); as "resembling *C. latrans*" (Wilson, 1933:68), as *C. latrans* (Anderson, 1968: 22).

Rancho La Brea, Los Angeles, Los Angeles County; Wisconsin; as *C. ochropus orcutti* (Merriam, 1912:255), as *C. latrans orcutti* (Giles, 1960); 60 skulls and 50 mandibles, LACM. Compared to those of Recent coyotes, Merriam stated that skulls of *orcutti* averaged larger, and had broader palates and zygomata, deeper and thicker mandibles, and thicker carnassials. Giles (1960) found no significant difference between samples of *orcutti* from Rancho La Brea and McKittrick. He also reported that the statistical separation among the Recent subspecies, *lestes*, *mearnsi*, and *ochropus*, was considerably less than that between *orcutti* and any one of them. My own findings concur with those of Merriam and Giles. The skulls of *orcutti* average larger in all measurable dimensions than those of Recent subspecies,

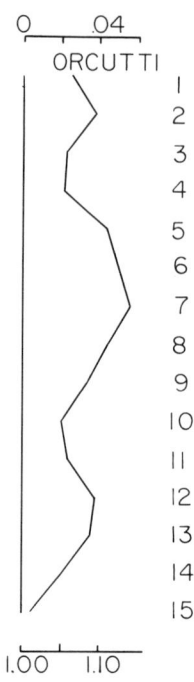

Fig. 48.—Ratio diagram comparing means of *C. latrans orcutti* from Rancho La Brea and the total series of Recent *C. latrans* from northern and western North America (vertical line). Vertically arranged numbers represent the measurements so numbered in appendix B. A log difference scale is provided above, and a ratio scale below the diagram. The Recent series consisted of the combined male and female samples (n=277). Sample sizes for the various measurements of Rancho La Brea coyotes were as indicated in appendix B (part 10).

and several individuals from Rancho La Brea are larger than the largest Recent specimens that I examined. Comparisons of proportion are provided in figures 44, 48, and 49. In addition, *orcutti* differs from Recent *C. latrans* in usually having more convex frontals, a more prominent sagittal crest (but not matching that of *C. lepophagus*), a broader supraoccipital shield, and temporal ridges that join anterior, rather than posterior, to the coronal suture. In size and all other characters there is overlap with series of living coyotes, and I agree with earlier authors in recognizing *orcutti* as a local late Pleistocene population of *C. latrans*. The fact that the coyotes in this group were large may be correlated with the sympatric presence of a much larger kind of wolf (*C. dirus*) than the kind (*C. lupus*) that shared the historical range of *C. latrans*. Hence the morphology and ecological niche of the late Pleistocene coyotes may have been shifted toward those of wolves. Merriam (1912:258) noted that a few cranial fragments from Rancho La Brea suggested the presence of a smaller coyote, more like that found presently in the area, in addition to *orcutti*. Schultz (1938b:164-168) also considered that two subspecies might be represented in the McKittrick deposits. But each of these authors decided that it would be best to refer all specimens of *C. latrans* from the respective sites to a single subspecies. I found one adult skull from Rancho La Brea, which was strikingly smaller than the others, but considering the variation that can be expected in a local population of Recent *Canis*, there is no valid reason why this single skull should be taxonomically distinguished from *C. l. orcutti*. And we would expect variation in the tar pit fauna to be especially great since the deposits were formed over thousands of years. Both in state of preservation and in abundance, the collections from Rancho La Brea contain the most useful series of Pleistocene *C. latrans*. The number of individual animals represented in the Los Angeles County Museum was counted at 200 by Stock (1929) and at 239 by Marcus (1960).

Rancho La Brea, Los Angeles, Los Angeles County; Wisconsin; as *C. andersoni* (Merriam, 1910: 393; 1912:260); skull without mandibles, UCMP 12249. Merriam specifically distinguished this specimen from *C. latrans* primarily because of its relatively short and broad rostrum. Jackson (1951:232) thought it possible that additional material might show *andersoni* to be synonymous with the living *C. latrans clepticus* of Baja and southern California. Giles (1960) also questioned the specific status of *andersoni*, and found statistical similarity between *andersoni* and four skulls of *clepticus* from San Diego County, California. Slaughter (1966a:479-481) found the rostral proportions of *andersoni* to be comparable to those of a Pleistocene skull from the Laubach Cave, Williamson County, Texas, which he thought might be referable to *C. latrans harriscrooki*. Anderson (1968:24) suggested that *andersoni* was probably a subspecies of *C. latrans*. The name *C. andersoni* here is synonymized under *C. latrans*, and is referred to the Rancho La Brea population of *C. l. orcutti*. The single skull upon which Merriam based his description, is from a juvenile, perhaps five months old at time of death. Although the rostrum is indeed relatively broader than in nearly all other available specimens of *C. latrans* (Fig. 49), this condition may be accounted for in part by age or by retarded development lengthwise. There are several other visible abnormalities that might contribute to the aberrant appearance of the skull. The alveolus for the left P1 is missing, and the alveolus for the right P1 crowds that for the right P2. The whole skull seems to be slightly twisted out of line, and there is an unusual inflation of the dorsal surface of the rostrum immediately anterior to the orbits. The characters of the occipital, braincase, frontals, and sagittal crest are those of normal *C. latrans*, and confirm that the specimen is a coyote, not a "coyote-like wolf" as stated by Merriam (1912:260).

Harbor freeway, Los Angeles, Los Angeles County; Wisconsin; as *C.* cf. *latrans* (Miller, 1971: 54).

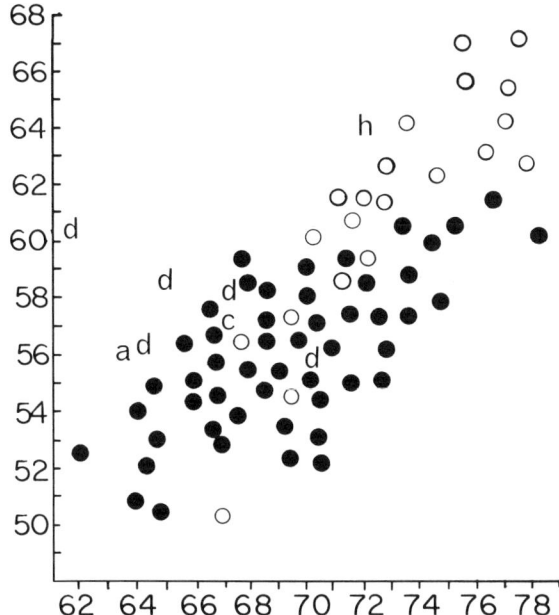

Fig. 49.—Scatter diagram comparing alveolar length of maxillary toothrow (horizontal axis) to maximum crown width across upper cheek teeth (vertical axis) in certain specimens of *C. latrans* (see appendix B for full description of measurements). Black dots, Recent *C. latrans lestes;* open circles, Pleistocene *C. latrans orcutti* from Rancho La Brea; c, *C. latrans caneloensis;* d, *C. latrans hondurensis;* h, *C. latrans harriscrooki;* a, *C. andersoni* (= *C. latrans orcutti*).

La Mirada, Los Angeles County; Wisconsin; as *C.* cf. *latrans* (Miller, 1971:49).

Costeau pit, 2 mi. S El Toro, Orange County; Wisconsin; as *C.* cf. *latrans* (Miller, 1971:17); M1, LACM 18220.

Vallecito Creek, San Diego County; Irvingtonian (Hibbard, *et al.*, 1965); six mandibular fragments, various postcranial fragments, LACM.

COLORADO.—Chimney Rock animal trap, Larimer County; late Pleistocene or early Recent; as *C. latrans* (Hager, 1972:65).

FLORIDA.—Ichetucknee River, Columbia County; Wisconsin; as *C. latrans* (Webb, 1974b:17). According to Kurten (1974:10) a left mandible (UF 1151) from this site represents a "medium to large" *C. latrans.* Martin and Webb (1974:128), however, stated that the same specimen "belongs to a smaller individual of the same general character" as material from Devil's Den, which they referred to *C. familiaris* (see below). In my own opinion UF 1151 represents *C. familiaris,* but Webb (1974b:17) may have based his report of *C. latrans* on other material. In my dissertation (Nowak, 1973:204) I associated another mandible (UF 11517) with the Ichetucknee River site, but that specimen actually is from Devil's Den.

Haile XIIB, Alachua County; Rancholabrean; mandibular fragment, UF.

Devil's Den, near Williston, Levy County; late Wisconsin or early Recent (7,000-8,000 B.P.); three mandibular fragments, UF 11514, 11515, 11517. Subsequent to my assignment of these specimens to *C. latrans* (Nowak, 1973:204-205), they, along with additional material from the site, were referred to *C. familiaris* by Martin and Webb (1974:127-128). The suggestion by these same authors, that canid specimens from the Ichetucknee River, Seminole Field, and Melbourne localities represent *C. familiaris,* rather than *C. latrans,* has raised questions about the over-all status of the coyote line in the Wisconsin and early Recent of Florida.

Reddick IA, Marion County; Sangamon (Webb, 1974b:13); as *C. latrans* (Gut and Ray, 1964:325).

Withlacoochee River VIIA, Citrus County; Sangamon; as *C. latrans* (Webb, 1974b:13, 17).

Seminole Field, near St. Petersburg, Pinellas County; Wisconsin (Hibbard, *et al.*, 1965); as *C.* cf. *riviveronis* (Simpson, 1929a:573), as *C. familiaris* (Martin and Webb, 1974:128).

Lake Cutaline, Pinellas County; late Pleistocene; mandibular fragment, UF.

Phillipi Creek-Fruitville Ditch, 7 mi. E Sarasota, Sarasota County; Wisconsin; as *C.* cf. *riviveronis* (Simpson, 1929b:275).

Melbourne, Brevard County; Wisconsin (Hibbard, *et al.*, 1965); as *C. riviveronis* (Gazin, 1950:12; Simpson, 1929b:268), as *C.* cf. *latrans* (Ray, 1958: 433), as *C. familiaris* (Martin and Webb, 1974: 128); rostral fragment, MCZ 5909; mandibular fragment, isolated teeth, USNM.

Vero (stratum 3), Indian River County; late Wisconsin (Webb, 1974b:13); as *C.* cf. *latrans* (Sellards, 1916:157), as *C. riviveronis* (Hay, 1917a: 59), as *C. latrans* (Weigel, 1962:38); maxilla with P4, FGS 7036. In his description of *C. riviveronis,* Hay wrote that the specimen differed from those of *C. latrans* in having a relatively shorter anterior lobe of P4, and a relatively greater transverse extent of the sockets of M1 and M2. The measurements he provided, however, were only slightly different from those of four comparative specimens of *C. latrans.* Ray (1958:433) considered that there was no basis upon which *C. riviveronis* could be distinguished from *C. latrans.* I agree, but the Vero specimen and others of *C. latrans* from Florida average smaller than western coyotes, possibly because of the sympatric presence of a small wolf (*C. rufus*).

IDAHO.—Jaguar Cave, Beaverhead Mountains, Lemhi County; late Wisconsin (C-14 dates: 10,370 ±350 and 11,580±250 B.P.); as *C. latrans* (Kurten and Anderson, 1972:24).

Moonshiner Cave, Bingham County; late Wisconsin or early Recent; as *C. latrans* (Kurten and Anderson, 1972:37).

Middle Butte Cave, Bingham County; early Recent; as *C. latrans* (Kurten, 1974:9).

American Falls, Power County; Rancholabrean (Hibbard, *et al.*, 1965), Illinoian (Kurten, 1974:7);

as *C. latrans* (Kurten, 1974:7). According to Kurten, two mandibles from this locality are "large." With respect to material from the same site, Gazin (1935:298) stated: "An incomplete humerus, a tibia and a third metatarsal are recognized as belonging to a dog somewhat smaller than *Canis occidentalis* but larger than a coyote." The possibility that all of this material represents the lineage of *C. rufus*, should not be overlooked.

Rainbow Beach local fauna, American Falls Reservoir, Power County; Wisconsin (C-14 dates: 21,500±700 and 31,300±2,300 B.P.); as *C. latrans* (McDonald and Anderson, 1975:26).

Twin Falls, Twin Falls County; Wisconsin; as *C. latrans* (Kurten, 1974:9).

ILLINOIS.—Galena, Jo Daviess County; Wisconsin (Kurten, 1974:10); as *C. latrans* (Hay, 1923:337).

Polecat Creek gravel pits, 1 mi. S Ashmore, Coles County; late Wisconsin (Hibbard, *et al.*, 1965); as *C. latrans* (Galbreath, 1938:306, 311).

INDIANA.—Boone County; Wisconsin (Kurten, 1974:10); as *C. latrans* (Cope and Wortman, 1884:7; Lyon, 1936:150).

IOWA.—Dubuque, Dubuque County; Wisconsin; as *C. latrans* (Kurten, 1974:9).

KANSAS.—Borchers local fauna, sec. 21, T33S R28W, Meade County; Yarmouthian; as *C.* cf. *latrans* (Getz, 1960:363).

Adams local fauna, north of Cimarron River, Meade County; early Illinoian (Hibbard, 1970); as *C. latrans* (Schultz, 1969:30).

Butler Spring local fauna, 15 mi. SSW Meade, Meade County; late Illinoian; as *C.* cf. *latrans* (Hibbard and Taylor, 1960:178).

Cragin Quarry local fauna, north of Cimarron River, Meade County; Sangamon; as *C.* cf. *latrans* (Hibbard and Taylor, 1960:178).

MARYLAND.—Cumberland Cave, 4 mi. NW Cumberland, Allegany County; Illinoian (Kurten, 1963:100); as *C.* cf. *priscolatrans* (Gidley and Gazin, 1938:23); cranial fragment, USNM 7660. The specimen is in poor condition, and represents an animal that may not have been a year old. The skull is the size of that of a large Recent *C. latrans*, and the braincase and postorbital constriction are coyotelike in shape. The frontals are more convex than in most Recent *C. latrans*.

MISSISSIPPI.—Vicksburg (south of), Warren County; "Wisconsin?"; as *C. latrans* (Kurten, 1974:10).

MISSOURI.—Brynjulfson Caves, 6 mi. SSE Columbia, Boone County; late Wisconsin (about 10,000 B.P.); as *C. latrans* (Parmalee and Oesch, 1972:29).

Younger's Cave, St. Clair County; early Recent; as *C. latrans* (Kurten, 1974:10); two mandibles, KU 5952, 7072.

Herculaneum (near), Jefferson County; Wisconsin (Hibbard, *et al.*, 1965); as *C. latrans* (Olson, 1940:42).

Bat Cave, 8 km. NW Waynesville, Pulaski County; late Wisconsin (10,000-16,000 B.P.); as *C. latrans* (Hawksley, Reynolds, and Foley, 1973:77).

Zoo Cave, 1 mi. ENE Hilda, Taney County; early Recent (less than 9,000 B.P.); as *C. latrans* (Hood and Hawksley, 1975:24; Saunders, 1977:14).

NEBRASKA.—Hay Springs quarry, Sheridan County; probably Illinoian (Hibbard, 1958); as *C.* cf. *latrans* (Matthew, 1902; Schultz, 1934:369).

Rushville fossil quarry, Sheridan County; Yarmouthian (Schultz and Martin, 1970); as "*Canis* sp.—Coyote" (Schultz and Tanner, 1957:71); mandibular fragment, UN 2913.

Mullen, Cherry County; late Irvingtonian (Kurten, 1974:7); as *C. latrans* (Martin, 1972:174); mandible, UN 26115.

Schmidt gravel pit, West Point, Cuming County; late Pleistocene; skull without mandibles, UN 2909.

Allen site, Frontier County; late Wisconsin; as *C. latrans* (Schultz, Martin, and Tanner, 1970:120).

Angus fossil quarry, Nuckolls County; Yarmouthian; as *C. latrans* (Schultz and Tanner, 1957:67).

NEVADA.—Tule Springs, Clark County; late Wisconsin; as *C. latrans* (Mawby, 1967).

NEW MEXICO.—Isleta Caves, 8 mi. W Isleta, Bernalillo County; late Wisconsin; as *C. latrans* (Harris and Findley, 1964:115).

Blackwater Draw, near Clovis, Curry County; Wisconsin (Lundelius, 1967:301); C1, M1, TM 937-896, 937-897.

Shelter Cave, near Las Cruces, Dona Ana County; late Pleistocene; skull without mandibles, LACM.

Conkling Cavern, near Las Cruces, Dona Ana County; late Pleistocene; skull, LACM 1634; mandible, LACM 1631.

Burnet Cave, 50 mi. W Carlsbad, Eddy County; Wisconsin (Hibbard, 1958); as *C. latrans lestes* and *C. microdon* (Schultz and Howard, 1935:284); mandibular fragment, C1, P4, UN 13454, 13455.

Dark Canyon Cave, Eddy County; late Pleistocene; mandible, LACM.

Dry Cave, 15 mi. W Carlsbad, Eddy County; Wisconsin (Kurten, 1974:8); as *C. latrans* (Harris, 1970:14).

OKLAHOMA.—Berends local fauna, near Gate, Beaver County; probably Illinoian (Hibbard and Taylor, 1960:57); as "a canid the size of a coyote" (Rinker and Hibbard, 1952:101), as *C. latrans* (Starrett, 1956:1187); mandibular fragment, UMMP 33319.

Afton, Ottawa County; Wisconsin (Kurten, 1974:9); as *C. latrans* (Hay, 1920:129); axis, USNM 9131.

OREGON.—Fossil Lake, Lake County; early or middle Wisconsin (Allison, 1966:32); as *C. lestes* (Elftman, 1931:7); three mandibular fragments, AMNH 8584, 8585, 8586.

PENNSYLVANIA.—Frankstown Cave, Blair County; Wisconsin (Hibbard, 1958); as *C. priscolatrans* (Peterson, 1926:283); two mandibular fragments (from same individual), C1, CM 11027. Peterson thought that the material indicated an animal about the same size as that represented by the type of *C. priscolatrans* Cope, 1899 (which in this paper is referred to a subspecies of *C. rufus*). The mandibles and teeth are large, but fall within the size range of Recent and Pleistocene *C. latrans* from western North America. As noted by Peterson, there is a prominent posterior cusp on the p2 of each mandible. In this respect, the Frankstown mandibles resemble those of *C. latrans harriscrooki* Slaughter, 1961 from the late Pleistocene of Texas.

TEXAS.—Rock Creek, Briscoe County; Kansan (Hibbard, 1970); as *C. priscolatrans* (Troxell, 1915: 628, 634). The description and illustration provided by Troxell indicate that an M1 from this site is best referred to *C. latrans,* not *C. rufus priscolatrans.*

Quitaque local fauna, Motley County; Wisconsin (Kurten, 1974:9); as *C. latrans* (Dalquest, 1964: 501).

Howard Ranch local fauna, Groesbeck Creek, northwest of Quanah, Hardeman County; Wisconsin (C-14 date: 16,775±565 B.P.); as *C. latrans* (Dalquest, 1965:71).

Wichita Falls, Wichita County; early Recent; as *C. latrans* (Dalquest, 1961:75).

Lubbock Reservoir, Lubbock County; Wisconsin (Lundelius, 1967:302); as *C. latrans* (Kurten, 1974: 9); mandibular fragment, TM.

Slaton quarry, 5 mi. N Slaton, Lubbock County; Sangamon (Hibbard, 1970); as near *C. latrans* (Dalquest, 1967:9).

Gilliland local fauna, Knox County; Irvingtonian (late Kansan: Hibbard, et al., 1965); as *C.* cf. *latrans* (Hibbard and Dalquest, 1966:20).

Benjamin Franklin local fauna, North Sulphur River, Delta County; late Wisconsin (11,000 B.P.); as *C. latrans* (Slaughter and Hoover, 1963:141). Two mandibular fragments were reported not to have a posterior cusp on p2, as typical of *C. latrans harriscrooki* (see following accounts). The material was thought to represent a more modern coyote that was either a replacement for or a descendent of the older *harriscrooki.*

Clear Creek local fauna, north of Denton, Denton County; Wisconsin (C-14 date: 28,840±4,740 B.P.; Hibbard, et al., 1965); as *C. latrans* cf. *harriscrooki* (Slaughter and Ritchie, 1963:125).

Lewisville site, Denton County; Sangamon or Wisconsin interstadial; as *C. latrans harriscrooki* (Slaughter, 1961); cast of type, mandible, SMUMP 60315. This subspecies originally was called "wolf like," and was distinguished from other coyotes by its well developed posterior cusp on p2, longer toothrow relative to depth of mandible, reduced distance between premolars, and more vertical ascending ramus. Slaughter examined 52 mandibles of Recent *C. latrans* from the United States, and found none with a posterior cusp on p2. He said, however, that this cusp was present on two specimens from San Luis Potosi, Mexico, and on one from Archaga, Honduras. The same condition existed in late Pleistocene specimens from Brazos County and Ingleside, Texas, which Slaughter referred to *harriscrooki*. Another mandible, collected at a Pleistocene site in Henderson County, Texas, also was assigned to this subspecies, on the basis of its more vertical ascending ramus. Slaughter speculated that *harriscrooki* might be a southern kind of coyote that could have inhabited Texas only in an interglacial or interstadial. Possible affinity to Recent *C. l. hondurensis* of Honduras was implied, and a jaw of that subspecies reportedly had the angle of ascending ramus about the same as in *harriscrooki.* I examined six specimens of *C. l. hondurensis* (1 in AMNH, 1 in KU, 2 in MCZ, 2 in USNM), of which five had a prominent posterior cusp on p2, and one had the cusp slightly developed. In contrast, only six of 250 Recent specimens from the western United States, and only one of 40 Pleistocene specimens from Rancho La Brea, which could be checked for this character, had any trace of the cusp. *Hondurensis* originally was reported to have a relatively broad palate, and I found such a condition to exist in most skulls of that subspecies that I examined (Fig. 49). Slaughter (1966b) reported that a Pleistocene skull from Laubach Cave, Williamson County, Texas, which he said might be referable to *harriscrooki,* had a relatively broader palate than Recent *C. latrans.* Therefore, it seems a reasonable hypothesis that a warmth-adapted coyote, with certain more wolflike characters than typical Recent *C. latrans,* was found in Texas during part of the Pleistocene, and might still be represented by the living coyote of Honduras.

Moore Pit local fauna, Dallas, Dallas County; Sangamon; as *C. latrans* cf. *harriscrooki* (Slaughter, 1966a:481; 1966b:79).

Trinity River terraces, 2.5 mi. NW Trinidad, Henderson County; Wisconsin; as "*Canis* sp.—Coyote" (Stovall and McAnulty, 1950:233), as *C. latrans harriscrooki* (Slaughter, 1961:509).

Clamp Cave, San Saba County; early Recent; as *C. latrans* (Lundelius, 1967:293).

Carson Holloway Ranch, San Saba County; Wisconsin; as *C. latrans* (Kurten, 1974:9).

Miller's Cave (Travertine unit), Llano County; early Recent; as *C. latrans* (Lundelius, 1967:293).

Longhorn Cavern, 8.5 mi. S Burnet, Burnet County; late Wisconsin (Hibbard, et al., 1965); as *Canis* (Semken, 1961), as *C. latrans* (Lundelius, 1967:293).

Laubach Cave, Georgetown, Williamson County; Wisconsin (Kurten, 1974:9); as *C. latrans,* possibly *C. l. harriscrooki* (Slaughter, 1966a:479-481); skull without mandibles, SMUMP 61269. As Slaughter reported, the specimen has a relatively broad facial width as compared to most *C. latrans.* Some Recent and Pleistocene specimens, however, especially those assigned to *C. latrans hondurensis,* approach the Laubach Cave specimen in relative broadness (Fig. 49). Nonetheless, assignment to *harriscrooki* seems a reasonable procedure (see account of Lewisville site, above).

Cage gravel pit, 5 mi. N Cameron, Milam County; late Pleistocene; "closely related to *C. latrans*" (Hay, 1927:291).

Brazos County; late Pleistocene; as "a coyote, *Canis* sp." (Peterson, 1946:166), as *C. latrans harriscrooki* (Slaughter, 1961:509).

Levi shelter, Travis County; early Recent; as *C. latrans* (Lundelius, 1967:293).

Schulze Cave, 28 mi. NE Rock Springs, Edwards County; Wisconsin or early Recent; "probably referable to *C. latrans harriscrooki* or . . . intermediate between that extinct race and the modern coyote" (Dalquest, Roth, and Judd, 1969:255-257).

Klein Cave, 12 mi. WSW Mountain Home, Kerr County; late Wisconsin; as *C. latrans* (Roth, 1972:78).

Cave Without a Name, Kendall County; late Wisconsin (C-14 date: 10,900±190 B.P.); as *C. latrans* (Lundelius, 1967:293).

Wunderlich site, Comal County; early Recent; as *C. latrans* (Lundelius, 1967:293).

Friesenhahn Cave, near Bulverde, Bexar County; Wisconsin; as *C. latrans* (Hay, 1920:141; Lundelius, 1960:38); four mandibles, TM 933-670, 933-1622, 933-2454, 933-3398. No posterior cusp was present on the p2 of the one specimen that could be checked for this character.

Ingleside gravel pit, San Patricio County; Wisconsin; as *C. latrans harriscrooki* (Slaughter, 1961:509); as *C. latrans* (Lundelius, 1972:20).

UTAH.—Silver Creek local fauna, 5 mi. N Park City, Summit County; late Sangamon to early Wisconsin; as "*Canis? latrans*" (Miller, 1976:401).

WISCONSIN.—Blue Mounds, Dane County (or "Iowa lead region"); late Pleistocene; as *C. latrans*. O. P. Hay apparently wrote of this same material three different times, stating (1) that it was probably from Iowa (1914:491); (2) that it was probably not from Blue Mounds, but from another crevice (1918:347); and (3) that it was found at Blue Mounds (1923:341).

WYOMING.—Little Box Elder Cave, west of Douglas, Converse County; Wisconsin; as *C. latrans* (Anderson, 1968, 24).

Bell Cave, Albany County; Wisconsin to early Recent; as *C. latrans* (Anderson, 1974:81).

AGUASCALIENTES.—Cedazo local fauna, near City of Aguascalientes; early Rancholabrean (probably Illinoian); as *C. latrans* (Mooser and Dalquest, 1975:786).

ESTADO DE MEXICO.—Tequixquiac (near); late Pleistocene; as "*C.* cf. *ocropus*" (Furlong, 1925: 139, 152), as "*C. ocrupus*" (Alvarez, 1965:27). Both of these authors apparently were referring to *C. ochropus,* a name for the living coyote of California, that has been arranged as a subspecies of *C. latrans.*

NUEVO LEON.—San Josecito Cave, near Aramberri; Wisconsin; as *C. latrans* (Kurten, 1974:7); 6 cranial fragments and 22 mandibles, LACM. The upper cranial elements can not be fully evaluated because of their poor condition, but seem not to differ from typical *C. latrans.* Measurements of the mandibles approach those of specimens from Rancho La Brea. None of the mandibles have a p2 with a posterior cusp. Age has not yet been reliably determined for the mammalian fauna of this site, but Jakway (1958:326) suggested that it was approximately as old as the fauna of Papago Springs Cave, Arizona (Wisconsin), and older than that of Rancho La Brea.

OAXACA.—Monte Flor Cave, 2 km. NE Valle Nacional; early Recent; as *C. latrans* (Alvarez, 1963).

PUEBLA.—Valsequillo, near Puebla; late Pleistocene; as *C. latrans* (Kurten, 1967:173).

Evolutionary position.—The species *C. latrans* apparently arose from certain populations within the species *C. lepophagus* in the Blancan. Subsequently, there appears to have been relatively little change in the coyote line, at least with respect to the skull, but some Pleistocene populations became larger and more massive than most, if not all Recent subspecies.

Canis edwardii Gazin

1942. *Canis edwardii* Gazin, Proc. U.S. Natl. Mus., 92:499.

1954. *Canis lupus baileyi,* Hoffmeister and Goodpaster, Illinois Biol. Monogr., 24:34.

Holotype.—Skull with mandibles; no. 12862, U.S. Natl. Mus.; about two miles northeast by east of Curtis Ranch House, San Pedro Valley, Cochise County, Arizona.

Geological distribution.—Late Blancan to early Irvingtonian.

Geographical distribution.—Known from Arizona, Kansas, Oregon, Texas, and Chihuahua.

Description.—A medium-sized canid resembling *C. rufus* in most observable characters; skull medium-sized with mostly narrow proportions; rostrum elongated and narrow; braincase relatively small and moderately inflated dorsoposteriorly; postorbital constriction elongated, broad lateromedially; zygomata slender, not deep; frontals moderately elevated above rostrum, not prominently convex; sagittal crest prominent; mandible long, slender, and shallow, with ascending ramus set at comparatively high angle to the vertical; teeth relatively large and set closely together in jaws; upper canines prominent, thin anteroposteriorly; P4 with moderately developed deuterocone; M1 with relatively large, deeply sculptured medial section, and pronounced buccal cingulum; p2 lacking posterior cusp; p3 usually with second cusp; p4 usually with second and third cusp, and pronounced posteromedial cingulum extending behind third cusp.

Comparison with C. lepophagus.—Larger and relatively broader in all measurable dimensions; teeth usually with less trenchant cusps.

Comparison with C. rufus.—Close resemblance in size and proportions of skull; postorbital constriction of braincase relatively broader; ascending ramus of mandible set at more acute angle to the vertical; M1 with broader medial section and more prominent buccal cingulum.

Remarks.—Hoffmeister and Goodpaster (1954:34) considered the name *C. edwardii* a synonym of *C. lupus baileyi*, the small gray wolf found in southern Arizona in historic time. Actually, this Pleistocene wolf can not be referred to a living subspecies. As is discussed later in this paper, *C. lupus* seems not even to have entered North America until the Illinoian, and the Curtis Ranch fauna is early Irvingtonian in age. Although, as explained by Hoffmeister and Goodpaster, there is little difference in over-all size between the type of *C. edwardii* and some specimens of *baileyi*, most measurable dimensions of *C. edwardii* are much smaller than the means of those of *baileyi*. Moreover, the rostrum of *C. edwardii* is narrower and the mandible shallower, than those of any skull of *baileyi* examined by me. The M1 of *C. edwardii* differs greatly from that of any modern *C. lupus*, in having a pronounced buccal cingulum and a relatively large, deeply sculptured medial section. In these and other features, the skull of *C. edwardii* approaches that of *C. rufus*, and, as realized by Gazin, that species is the only one to which the fossil need be critically compared. Indeed, the red wolf probably is a direct descendent or immediate relative of *C. edwardii*, and the latter eventually may be shown to be only a synonym or subspecies of *C. rufus*.

In my dissertation (Nowak, 1973:208, 229-230), I had indicated that *C. edwardii* was known only from the type locality, but that several other early wolf specimens might be referable to this species, rather than to *C. rufus*. Additional material, subsequently made available to me through the kindness of Richard H. Tedford and Beryl E. Taylor at the American Museum of Natural History, suggests a greater range for *C. edwardii* and also provides a basis for assigning some previously examined specimens to this species. Of particular importance in this regard is a skull from the Rome Beds, Oregon, which allows direct comparison between its own parts and those of material from the type locality and from the Gilliland local fauna, Texas. Cranial elements from the latter site appear to have about the same size and shape as those of the Rome Beds specimen. The type of *C. edwardii*, as well as a maxillary fragment from the same locality, not mentioned in Gazin's (1942) original description, have an M1 with a remarkably pronounced buccal cingulum, and the Rome Beds specimen shares this character. This cingulum is never so prominent in *C. rufus* as in these three specimens, and this factor is one reason for not now synonymizing *C. edwardii* with the red wolf.

Kurten (1974) considered the type of *C. edwardii* to be one of the large Irvingtonian and late Blancan coyotes which he grouped under the name *C. priscolatrans* (see p. 75). Although Kurten correctly associated *C. edwardii* with *C. priscolatrans*, I think that both names represent the lineage of the wolf, rather than the coyote.

Record of occurrences.—The following list is arranged alphabetically by state, except that Chihuahua is placed last. Specimens examined by me are identified by element, museum number, or both; and selected measurements are found in appendix B (part 11) and appendix C (part 3). Occurrences also are shown on the map in figure 50.

ARIZONA.—Anita, Coconino County; early Irvingtonian (Richard H. Tedford, American Museum of Natural History, pers. comm.); as *"C. nubilus?"* (Hay, 1921:632); two mandibular fragments, USNM 10210 B and C. As reported by Hay, the most complete ramus, in comparison with that of Recent *C. lupus*, is lower and thinner, and has thinner teeth. The proportion of length to depth actually is unlike that found in any specimen of *C. lupus*, but is close to that in some specimens of Recent *C. rufus*. The other ramus also is shallow, but is not complete enough for full evaluation. Still another mandibular fragment was found at the site, and was questionably referred to *C. lupus* by Hay. This specimen is much larger and deeper than the other two, and probably represents *C. armbrusteri*. Its presence at Anita, together with the other specimens, suggests that by the early Irvingtonian there already had been a divergence between the line of small primitive wolves (*C. edwardii* and *C. rufus*) and the line leading to the larger wolves of the late Quaternary. Assignment of the Anita material to *C. edwardii*, rather than to *C. rufus*, is arbitrary because adequate samples are unavailable.

Curtis Ranch, San Pedro Valley, Cochise County; early Irvingtonian (Johnson, Opdyke, and Lindsay, 1975); as *C. edwardii* (Gazin, 1942:499); skull with mandibles, USNM 12862; maxillary fragment with P3-M2, USNM 12864.

KANSAS.—Arkalon gravel pit, south side of Cimarron River, Seward County; Irvingtonian (late Kansan); "The humerus, femur and other elements are the length of those of a large *Canis latrans* Say. The bones are heavy and nearly as large in diameter as those of *Canis lupus* Linne (Hibbard, 1953:115).

Cudahy fauna, Big Springs Ranch, Meade County; Irvingtonian (late Kansan: Hibbard, *et al.*, 1965); as *Canis* sp. (Getz, 1960:361). An astragalus was reported to be from a canid the size of a small wolf and larger than a coyote.

OREGON.—Rome Beds, Malheur County; Irvingtonian; as *C. priscolatrans* (Kurten, 1974:6); partial skull without mandibles, USNM 23898 (in U.S. Geological Survey collections, Menlo Park, California). The specimen is approximately the same size as the type of *C. edwardii*. The teeth are larger than those of nearly all coyotes examined.

TEXAS.—Gilliland local fauna, Knox County; Irvingtonian (late Kansan: Hibbard, *et al.*, 1965); as *C.* cf. *lupus* (Hibbard and Dalquest, 1966:20); rostral fragment, UMMP 46483; parietal, UMMP 46460. Hibbard and Dalquest thought these fragments to represent "a canid the size of the gray wolf," but actually the specimens are smaller than the corresponding parts of any skull of *C. lupus* examined by me. They clearly are not referable to *C. latrans* or *C. lepophagus*, and in features that can be evaluated they closely resemble the specimen of *C. edwardii* from Rome Beds, and also fall well within the range of variation of *C. rufus*. Hibbard (pers. comm.) came to consider the Gilliland fauna to be pre-Kansan, and, if so, these specimens represent one of the earliest known occurrences of a wolf in North America.

CHIHUAHUA.—Miña Erupcion, 90 mi. SSE Juarez; Pleistocene; as *Canis* sp. (Eaton, 1923:233). Eaton wrote that six vertebrae from an adult animal were smaller than those of *C. lupus*, but larger than those of *C. latrans*. He implied affinity to *C. priscolatrans* Cope, 1899.

Miñaca Mesa, approximately 100 mi. W, 10 mi. S City of Chihuahua; Blancan (Kurten, 1974:6); mandibular fragment, LACM 105/149. Although Kurten associated this specimen with the coyote line, it is larger than any mandible of *C. lepophagus* or *C. latrans* examined by me. The specimen does, however, resemble most available material of *C. lepophagus* in the pronounced development of the second cusp of p4, and the reduced development of the third cusp. In addition, unlike most wolves, p3 (as well as p2) lacks posterior cusps. While referral to *C. edwardii* still seems most appropriate, the possibility remains that this specimen represents a transitional phase through which the wolf line evolved from *C. lepophagus*.

Evolutionary position.—*Canis edwardii* may represent the first unquestionable appearance of a wolf in North America. Material referable to this species is on the average older and more primitive in characters than that assigned to other Pleistocene species of wolves. Perhaps *C. edwardii* descended from a late Blancan population of *C. lepophagus*, but we also can not rule out the alternative that the wolf and coyote lines had been distinct at an earlier time. The presence of a large wolf (*C. armbrusteri*) in the early Irvingtonian, as well as *C. edwardii* and *C. rufus*, suggests that radiation of the wolf group had been in progress for a considerable period. This radiation may have been associated with the initial glacial advances of the Pleistocene, and also with the simultaneous extinction of the large borophagine dogs.

The early history of the wolves, and their exact relationships with the coyotes, can not now be assessed because of the scarcity of fossil *Canis* in the Blancan and Hemphillian (middle Pliocene). A single mandibular fragment (UN 2908) from the Hemphillian

Mitchell Creek Ash Hollow formation, Frontier County, Nebraska, has measurable dimensions close to those of *C. edwardii*, and appears to represent a wolf. Because of the poor condition of this specimen, and its removal in time from the scope of this paper, I do not now refer it to a particular species.

Canis etruscus, a wolf resembling *C. edwardii*, was present in the early Pleistocene of Europe (Kurten, 1968:109). Thus, a group of small, relatively unspecialized wolves, retaining some coyotelike characters, seems to have become widespread at this time. This group apparently formed the basic stock from which the larger wolves of the late Quaternary descended. Kurten (1968:108-109) thought that *C. etruscus* probably gave rise to *C. lupus mosbachensis* of the middle Pleistocene, from which in turn modern *C. lupus* developed. In the New World, *C. edwardii* seems to have been close to the line from which arose the larger *C. armbrusteri* and *C. dirus*.

The geographical distribution of *C. edwardii* (Fig. 50) appears to have been concentrated in the southwestern quarter of the continent, while the closely related *C. rufus* occupied the southeast. The latter was able to survive, but *C. edwardii* eventually disappeared, perhaps because of changing habitat conditions or competition with *C. latrans* and/or *C. lupus*.

Canis rufus Audubon and Bachman

1791. *Lupus niger* Bartram, Travels, p. 199. Not available because Bartram was not consistently binomial (according to Int. Comm. Zool. Nomen., 1957, opinion 447).

1851. *Canis lupus* var. *Rufus* Audubon and Bachman, Quadrupeds of North America, 2:240. Type locality, 15 mi. W Austin, Texas (Goldman, Jour. Mamm., 18:38, 1937).

1899. *Canis priscolatrans* Cope, Jour. Acad. Nat. Sci., Philadelphia, ser. 2, 9:227. Type from Port Kennedy deposit, Upper Merion Township, Montgomery County, Pennsylvania. Valid as a subspecies of *C. rufus*.

1905. *Canis rufus*, Bailey, N. Amer. Fauna, 25:174.

1912. *Canis floridanus* Miller, Proc. Biol. Soc. Washington, 25:95. Type from Horse Landing, about 12 mi. S Palatka, Putnam County, Florida. Valid as a subspecies of *C. rufus*.

1937. *Canis rufus gregoryi* Goldman, Jour. Mamm., 18:44. Type from Macks Bayou, 3 mi. E Tensas River, 18 mi. SW Tallulah, Madison Parish, Louisiana.

1942. *Canis niger,* Harper, Jour. Mamm., 23:339.

1965. *Canis rufus*, Hall, Univ. Kansas Mus. Nat. Hist. Misc. Publ., no. 43, p. 13.

Recent subspecies revised by Goldman (1944); subspecies listed and distribution mapped by Hall and Kelson (1959:851-852), and by this paper (Fig. 50).

Type.—None designated.

Geological distribution.—Early Irvingtonian to Recent.

Geographical distribution.—Pleistocene and early Recent records from Arkansas, Florida, Pennsylvania, Texas, and Estado de Mexico; historical range confined to southeastern quarter of North America, from central Texas to Atlantic, and from Gulf of Mexico north to southern Pennsylvania, Ohio Valley, and southeastern Kansas; presently found only in extreme southeastern Texas and southern Louisiana.

Description.—Medium-sized for the genus; skull medium-sized with narrow proportions; rostrum elongated and narrow; braincase relatively small and not much inflated dorsoposteriorly; postorbital constriction elongated, narrow lateromedially, lateral margins often appearing parallel when viewed from above; zygomata usually slender and not widely spreading; orbits usually large; frontals usually moderately elevated above rostrum, not prominently convex, and forming a relatively narrow and flat shield; temporal ridges often sharp, often obscuring frontal suture, and usually joining anterior to coronal suture; sagittal crest prominent and sharp dorsally; supraoccipital shield moderately large; external side of occipital often well ossified; tympanic bullae usually well inflated; mandible long, narrow, and shallow, ventral edge usually not notably convex when viewed from side; incisors often relatively small; upper canines prominent, thin anteroposteriorly, ventral tips usually extending below level of anterior mental foramina when jaws are closed; premolars with trenchant, laterally compressed cusps; P4 usually with prominent deuterocone and lingual cingulum; M1 often having relatively large, deeply sculptured medial section, the metaconule usually prominent and well separated from protocone, buccal and anterior cingula usually pronounced; M2 relatively large, cusps well developed; p2 occasionally with a posterior cusp; p3 sometimes with second and third cusp; p4 with second cusp, usually with a moderately developed third cusp and posteromedial cingulum extending behind third cusp; m2 and talonid of m1 relatively large, with moderately trenchant cusps. For details on pelage and postcranial skeleton, see Goldman (1944), Young (1946:36), and Paradiso and Nowak (1972b).

Comparison with C. latrans.—Skull larger and relatively broader in most dimensions; rostrum usually relatively broader and deeper; braincase relatively smaller and not so much inflated, never broader at level of parietotemporal sutures than at base; postorbital constriction narrower and more elongated; zygomata deeper and more widely spreading; jugal more deeply inserted in maxilla; frontals often more elevated above rostrum; temporal ridges usually sharper, more often obscuring frontal suture, and usually joining anterior, rather than posterior, to coronal suture; sagittal crest more prominent; supraoccipital shield broader, projecting farther posteriorly; external side of occipital more ossified, more often lacking projection dorsal to foramen magnum; occipital condyles usually extending farther transversely; mandible usually relatively thicker and deeper; premolars set more closely together in jaws; upper canines thicker anteroposteriorly, usually not extending so far ventrally; premolars usually relatively broader with less trenchant cusps; deuterocone and lingual cingulum of P4 usually less prominent; M1 usually with relatively smaller medial section, and less prominent hypocone, metaconule, and buccal cingulum; p2 more often with posterior cusp; posteromedial cingulum on p4 usually less prominent; metaconid of m1 less pronounced, not projecting so far medially; m2 and talonid of m1 with less trenchant cusps. Atkins and Dillon (1971) listed differences between *C. rufus* and *C. latrans* in the morphology of the cerebellum; Russell and Shaw (1972) and Jackson (1951:240) discussed distinguishing characters in external appearance.

Comparison with C. lupus.—Cranial differences usually or more often apparent are as follows: skull smaller and relatively narrower in most dimensions; rostrum narrower; braincase relatively deeper; lateral margins of postorbital constriction appearing more nearly parallel when viewed from above, not rising so steeply into frontal region; zygomata more slender, not so deep, not so widely spreading; orbits relatively larger; frontals less elevated above rostrum, less convex, and forming a flatter and relatively narrower shield; tympanic bullae more inflated; mandible shallower; incisors smaller; upper canines thinner anteroposteriorly, extending more ventrally; premolars narrower; P4 with more prominent deuterocone, its root appearing to pass more vertically into palate; M1 with relatively larger, more deeply sculptured medial section, and more prominent metaconule and buccal cingulum; M2 relatively larger; p4 more often with third cusp and posteromedial cingulum extending behind third cusp;

talonid of m1 relatively larger with more trenchant cusps. Atkins and Dillon (1971) listed differences between *C. rufus* and *C. lupus* in the morphology of the cerebellum; Goldman (1944), and Paradiso and Nowak (1972b) discussed differences in pelage and external appearance.

Other comparisons.—See accounts of *C. edwardii* and *C. armbrusteri*.

Remarks.—Two critical problems concerning the systematics of *C. rufus* are: (1) its original relationship with *C. lupus*, particularly the question of whether the two intergraded in the forests of the eastern United States; and (2) the relationship in historic time between *C. rufus* and *C. latrans* in the south-central United States.

Although it is sometimes difficult to separate specimens of red and gray wolves, the previous part of this paper showed that multivariate analysis could distinguish nearly all skulls of *C. rufus*, including all taken before 1920 in the eastern United States, from large series of *C. lupus*. In addition to the measurements used in multivariate analysis, the characters listed in the above "comparison with *C. lupus*" usually serve to distinguish the two species. Problems may still arise, especially if complete skulls are not available.

In nearly all measurements and other features in which *C. rufus* differs from *C. lupus*, the former approaches *C. latrans*. Indeed, available specimens of the red wolf almost bridge the morphological gap between the proximal extremes of the other two species. Hybrid origin for *C. rufus* thus seems to be one possibility, but there are other solutions to the problem. The most reasonable explanation is that *C. rufus* represents a primitive line of wolves that has undergone less change than *C. lupus*, and has thus retained more characters found in the ancestral stock from which both wolves and coyotes arose.

The frontal shield and postorbital constriction are the only parts of the red wolf's skull that often do not have a form intermediate between that of typical *C. latrans* and *C. lupus*. The postorbital constriction in *C. rufus* is sometimes relatively narrower than in both the gray wolf and coyote, and the elongated lateral margins often appear parallel when viewed from above, unlike the normal condition in either of the other two species. The frontal shield of *C. rufus* is also relatively narrower, and in some specimens has a more flattened aspect than in either *C. lupus* or *C. latrans*. These characters, together with a prominently rising sagittal crest, give a unique appearance to certain specimens of *C. rufus*, including both some pre-1920 and some post-1960 individuals.

Lawrence and Bossert (1967) considered that if initial study of the red wolf had been based on adequate series from the southeastern United States (rather than from Texas), *C. rufus* and *C. lupus* probably would not have been taxonomically separated. As we have seen, however (pp. 25-28), complete skulls taken prior to 1920 in Louisiana and eastward are rare, and can all be distinguished from those of *C. lupus*. In addition to the 14 skulls listed in table 2, Goldman (1944) assigned two other early specimens to *C. rufus*. One of these, a subadult female taken in 1832 on the Wabash River, Indiana (in AMNH), was assigned to *C. rufus gregoryi*. I agree with this designation, as the specimen is comparatively small and narrow-proportioned, and has the dental characters normally associated with the red wolf. The other skull (in USNM), the type of *C. rufus floridanus*, was taken in 1890 on the St. Johns River, Putnam County, Florida. This specimen is difficult to evaluate because the posterior part is missing, and there is also a dental anomaly in that M2 on both sides is missing. Nonetheless, the specimen seems to be within the morphological range of other skulls of *C. rufus*.

Archeological sites and other Recent deposits in the eastern United States have yielded various specimens of wolves, but few in good condition. Such specimens examined by me, which seem best referred to *C. rufus,* include the following.

Banks site, 1.5 mi. N Clarksdale, Crittenden County, Arkansas; as *C. lupus* (Parmalee, 1959c:6); mandible, ISM.

Blain site, west bank Scioto River, south of Chillicothe, Ross County, Ohio; as *C. lupus* (Parmalee and Shane, 1970:198); maxillary fragment, ISM.

New Paris Sinkhole No. 2, Bedford County, Pennsylvania; as *C. lupus lycaon* (Guilday and Bender, 1958:134); incomplete skull and pair of mandibles, CM.

Eschelman site, 3 mi. S Columbia, Lancaster County, Pennsylvania; as *C. lupus* (Guilday, Parmalee, and Tanner, 1962:64); cranial fragment and pair of mandibles, CM.

Buffalo Village site, Putnam County, West Virginia; as *C. lupus* (Guilday, 1971:9); three mandibular fragments and isolated m1, CM.

Lauderdale Indian mound, Washington County, Virginia; isolated M1 (collection of Ronald M. Nowak).

Crow Island Indian midden, Jackson County, Alabama; mandible, UMMZ. Recently, Barkalow (1976: 25-26) reported this specimen, and material from two other sites in the vicinity, to be either *C. rufus* or *C. lupus.*

Jungerman site, Indian River, just south of southern tip of Merritt Island, Brevard County, Florida; as *C.* cf. *niger* (Wing, 1963:52); m1, m2, UF.

Nichol's Hammock, .7 mi. NE Princeton, Dade County, Florida; sinkhole with contemporary fauna; as *C. niger* (Hirschfeld, 1968:180); mandible, UF.

Bullen and Benson (1967) reported the discovery of three cut and perforated canid jaws on Tick Island, Florida. Although referral to *C. rufus* is a possibility, the fragmentary nature of the material (as illustrated by Bullen and Benson) would make assignment to any particular species of *Canis* difficult.

Webb and Baby (1957:61-71) described three specimens of wolves from the Wright Mounds, Montgomery County, Kentucky; near New Liberty, Owen County, Kentucky; and the Wolford Mounds, Pickaway County, Ohio. The most complete of these specimens, from the Wright Mounds, was a spatula-shaped artifact cut from the upper jaws and palate of a wolf. The specimen now has apparently been lost, but a published photograph suggests that the skull represented had a narrow rostrum, and may have belonged to a red wolf.

Fossil record.—The following list is arranged alphabetically by state, except that Estado de Mexico is placed last. Specimens examined by me are identified by element, museum number, or both; and selected measurements are found in appendix B (part 12) and appendix C (part 4). Occurrences also are shown on the map in figure 50.

ARKANSAS.—Eddy Bluff shelter, near Springdale, Washington County; early Recent; as *C. rufus* (Morrison, 1970); maxillary fragment, UArk. The fossil closely matches series of modern *C. rufus* in size and other characters.

FLORIDA.—Haile VIIA, Alachua County; Sangamon (Webb, 1974b:13); cranial fragment, UF. The specimen is larger than comparative material of *C. latrans* and is smaller than *C. lupus* or *C. dirus.* In size, and in characters of the frontal region and dentition, the specimen is well within the range of variation of Recent *C. rufus.* Martin (1974:77) compared measurements of the P4 (incorrectly labeled as p4 in his figure 3.13) of nine specimens from this site, to those of other wolves, and stated that either *C. lupus* or *C. rufus* was represented. According to his scatter diagram, the measurements of the Haile material are substantially closer to those of *C. rufus* than to those of *C. lupus.*

Devil's Den, near Williston, Levy County; late Wisconsin or early Recent (7,000-8,000 B.P.); as *C. rufus* (Martin and Webb, 1974:126).

Inglis IA, Citrus County; early Irvingtonian; as *C.* cf. *niger* (Klein, 1971:17), as *C. rufus* (Webb, 1974b:17), as *C. lupus* (Martin, 1974:72), as *C. priscolatrans* (Kurten, 1974:6); right and left maxillary fragments, UF 18046; P4, UF 18049; c1, UF 18052; M1, UF 19406; m1, UF 19404; two mandibular fragments, UF 19323, 19324. This material indicates the presence of a canid close in size and dental characters to the wolf (*C. rufus*) that inhabited the southeast in historical time. Several of the specimens are larger than those of any Recent or late Pleistocene coyote examined by me. The first upper molars in this series have deeply sculptured medial sections, as does the type of *C. priscolatrans,* but neither the Inglis nor Port Kennedy specimens have the buccal cingulum on M1 as strongly developed as in *C. edwardii.* Klein (1971:17-18) observed that the measurements of the Inglis specimens approached those of *C. edwardii* from Curtis Ranch, and that these specimens indicated a wolf very close to, if not conspecific with, *C. rufus.*

Crystal River Power Plant, Citrus County; Sangamon; maxillary fragment, UF 17074. Kurten (1974:10) assigned another maxilla from this site to *C. latrans,* but noted that it was "large." Possibly that specimen should be referred to *C. rufus.*

Melbourne, Brevard County; Wisconsin (Hibbard, *et al.,* 1965); as *C.* cf. *lupus* (Ray, 1958:434), as *C. rufus* (Webb, 1974b:17); mandibular fragment, MCZ 17789. As explained by Ray, *C. lupus* and *C. rufus* can not always be distinguished on the

Fig. 50.—Map showing localities of *C. rufus* from archeological sites (triangles), fossil *C. rufus* (black dots), and *C. edwardii* (squares). The solid lines show the distribution of Recent subspecies: *C. rufus rufus* (R), *C. rufus gregoryi* (G), and *C. rufus floridanus* (F). Because of the scale of the map, it was not possible to plot all localities in crowded areas.

basis of the mandible and lower teeth. I tentatively refer the Melbourne specimen to *C. rufus* because it is within the size range of that species, and because it has a posteromedial cingulum extending behind the third cusp of p4.

Vero, Indian River Country; late Wisconsin

(Webb, 1974b:13); as *C.* cf. *niger* (Weigel, 1962: 37). Only a P1 and the anterior half of a P4 were so referred, and Weigel considered the material too meager for positive identification. Webb (1974b:17), however, listed *C. rufus* for this site.

PENNSYLVANIA.—Port Kennedy deposit, Upper Merion Township, Montgomery County; Irvingtonian (probably Yarmouthian: Hibbard, 1958); as *C. priscolatrans* (Cope, 1899:227); P4, M1, M2, p4, ANSP 57-58. Cope regarded the upper teeth as "the type of a distinct species, having important points of resemblance to the coyote" (although his publication listed the premolar as a P1, the tooth actually is a P4). Cope also noted: "The forms of the cusps and cingula in this species are like those of the corresponding teeth of the coyote, except as to the conules. The size is that of the large, but not largest wolves." This description agrees well with that of *C. rufus* by Goldman (1944), and Paradiso and Nowak (1972a), who considered the red wolf to have coyotelike teeth, but to approach *C. lupus* in size. A very few specimens of Recent and late Pleistocene *C. latrans* have teeth as large as those found at Port Kennedy, but I disagree with Kurten's (1974) suggestion that the type of *C. priscolatrans* represents a coyote ancestral to modern *C. latrans*. In contrast, the Port Kennedy teeth fall well within the range of variation of Recent *C. rufus*, and are nearly equal in size to those of *C. edwardii*.

TEXAS.—Miller's Cave, Llano County; late Wisconsin; as *Canis* sp. (Patton, 1963:31). According to Patton, a single m2 from the site was larger than those of coyotes, and slightly larger than that of one available specimen of *C. rufus*.

Buffalo Bayou, Houston, Harris County; late Pleistocene; as "*Canis* sp. cf. *lupius* [sic] *baileyi*" (Du Bar and Clopine, 1961:99). Since the gray wolf is not known to have occurred near Houston in Recent time, since the nearest geographical subspecies to Houston is not *baileyi*, and since the specific name was not correctly spelled, there is reason to suspect that identification of the pertinent material was not made carefully. The specimen apparently has been lost, but I think that any late Pleistocene remains of small wolves in the area would be referable to *C. rufus*.

ESTADO DE MEXICO.—Upper Becerra formation, northwest of Puente del Gallo, Valley of Tequixquiac; Sangamon or Wisconsin; as *Canis* sp. (Hibbard, 1955:52). According to Hibbard, a mandible from the site "is smaller than *Canis lupus* Linnaeus and appears closely related to *Canis niger* (Bartram)."

Evolutionary position.—Modern *C. rufus* apparently represents a comparatively unmodified surviving line of the primitive stock of small wolves that had developed by the early Pleistocene. The red wolf evolved from *C. edwardii*, or a close relative, and then remained in North America through the middle and late Quaternary. The gray wolf probably evolved from a branch of the same stock, but one that had entered the Old World and become isolated there through factors associated with glaciation. While *C. lupus* developed in Eurasia and eventually became the only species of wolf throughout most of the Northern Hemisphere, and while *C. dirus* underwent its sudden rise and fall in the New World, the smaller *C. rufus* held on to its niche in the southern forests and marshes.

Goldman (1944:399) wrote that certain Pleistocene remains from Rancho La Brea suggested the presence there of a species with relationship to *C. rufus*. All of the specimens of wild *Canis* from Rancho La Brea that I examined, however, could be referred to *C. dirus, C. lupus,* or *C. latrans*.

There has been a suggestion that *C. rufus* evolved from a coyotelike ancestor that had become isolated by glaciation in a Florida refugium (Nowak, 1970:84). This hypothesis no longer is tenable in the light of the above outlined evolutionary sequence of the red wolf. Furthermore, the subspecies of coyote (*C. latrans riviveronis*) that inhabited Florida was small, and survived into early Recent time, and hence could not have given rise to the much larger *C. rufus* which already was present in Florida by the Irvingtonian.

Canis armbrusteri Gidley

1913. *Canis armbrusteri* Gidley, Proc. U.S. Natl. Mus., 46:98.

Type.—Portion of a left lower jaw containing p4 to m2; no. 7662, U.S. Natl. Mus.; Cumberland Cave, about 4 mi. NW Cumberland, Allegany County, Maryland.

Geological distribution.—Early (?) Irvingtonian to early Rancholabrean.

Geographical distribution.—Known from Maryland and Florida, with possible records from Arizona, California, Nebraska, Pennsylvania, South Carolina, and Texas.

Description.—Size large for the genus; skull usually large and relatively narrow in most proportions; rostrum elongated and narrow; braincase moderately inflated dorsally; zygomata usually deep and broadly spreading; frontals moderately elevated above rostrum, not prominently convex, and forming relatively narrow shield; sagittal crest prominent and sharp dorsally; supraoccipital shield large; tympanic bullae notably large and well inflated; mandible long, moderate in depth; teeth comparatively large; P4 usually with prominent deuterocone; M1 having relatively large, deeply sculptured medial section, and pronounced buccal cingulum; M2 relatively large; p2 and p3 lacking posterior cusps in available specimens; p4 having second and third cusps, and pronounced posteromedial cingulum extending well behind third cusp; talonid of m1 relatively large.

Comparison with C. rufus.—Usually much larger; postorbital constriction with lateral margins not parallel; zygomata usually deeper and more broadly flaring; mandible relatively deeper; teeth, especially carnassials, sometimes relatively larger; p4 having more pronounced posteromedial cingulum.

Comparison with C. lupus.—Skull usually narrower in most proportions; rostrum relatively longer and narrower; braincase usually more inflated dorsally; frontals less convex and usually forming relatively narrower shield; tympanic bullae larger and more inflated; P4 usually having more prominent deuterocone; M1 having relatively larger, more deeply sculptured medial section, and pronounced buccal cingulum; M2 relatively larger; p4 with third cusp, and pronounced posteromedial cingulum extending behind third cusp; m1 with relatively larger talonid.

Comparison with C. dirus.—Usually smaller; skull narrower in most proportions; rostrum relatively longer and much narrower; braincase more inflated dorsally; postorbital constriction not rising so steeply into frontal region; frontal shield much narrower; sagittal crest usually less prominent; supraoccipital shield broader and not projecting so far posteriorly; tympanic bullae larger and more inflated; postpalatine foramina more anteriorly placed (arrangement of the optic and anterior lacerated foramina can not be evaluated in available specimens of *C. armbrusteri*); anterior parts of vertical plates of palatines flaring less broadly (placement of the vomer can not be evaluated); mandible usually shallower; P4 usually relatively smaller with more prominent deuterocone; M1 having relatively larger, more deeply sculptured medial section, more reduced paracone and metacone, more prominent hypocone with its anterior ridge extending around protocone, and more pronounced buccal cingulum; M2 relatively larger; p4 usually similar; m1 smaller with relatively larger talonid.

Remarks.—Gidley's (1913) original description of *C. armbrusteri* was based on three lower jaws from Cumberland Cave, which reportedly differed from those of *C. lupus* in having relatively greater depth, smaller canines, p2 and p3 without posterior cusps, p4 with a third cusp and posterior cingulum, and m1 with a larger heel. Not all of these characters can now be considered diagnostic, but on the whole Gidley's distinction of *C. armbrusteri* was borne out by the discovery of additional material. The upper teeth of the species were first described by Patterson (1932), who, like Gidley, noted certain coyotelike characters. Gidley and Gazin (1938: 15-23) discussed a number of skulls and mandibles from Cumberland Cave, which supported the continued recognition of *C. armbrusteri*.

Martin (1974:76) suggested that *C. armbrusteri* is synonymous with *C. lupus,* and that specimens from Cumberland Cave are closely matched by skulls of large, northern gray wolves. I disagree with this interpretation for reasons provided in the above comparison of the two species. Probably of greatest value in distinguishing *C. armbrusteri* from Recent *C. lupus* is the presence in the former of a pronounced buccal cingulum on the M1, and a posteromedial cingulum extending well behind the third cusp on p4.

These characters are conspicuous in specimens both from Cumberland Cave and Florida. Also, the bullae of *C. armbrusteri* are more inflated than those of *C. lupus,* and, in fact, are larger than those of any other species of *Canis.*

Goldman (1944:399) thought that the Cumberland Cave wolf appeared closely allied to the red wolf. In most proportions and dental characters, *C. armbrusteri* does approach Recent *C. rufus,* and there would be a basis for considering it a giant Pleistocene red wolf. The two species are easily distinguished by size and other characters, however, and there is some evidence that they occurred together at certain Pleistocene localities.

Whereas some specimens of *C. armbrusteri* seem not very different from *C. rufus,* others are nearly as large as *C. dirus.* Although the range of variation shown by *C. armbrusteri* is not unusually great for a species of *Canis,* the available material does span much of the morphological gap between *C. rufus* and *C. dirus,* and may represent a part of the evolutionary sequence through which the dire wolf developed from more primitive stock. Martin (1974:75) reported that one of the Cumberland Cave skulls (USNM 11886) has an inion projection as pronounced and hooked as in *C. dirus,* and might represent a population beginning to develop into that species. In my own opinion this specimen does match some skulls of *C. dirus* in size and height of sagittal crest, but not in projection of inion. Furthermore, in all distinguishing characters of the dentition that can be evaluated, the specimen is unlike *C. dirus.*

Record of occurrences.—Only the remains from Cumberland Cave and two Florida sites are complete enough for reliable assignment to *C. armbrusteri.* Various other fragments of large wolves have been reported from pre-Illinoian sites, and are not definitely referable to *C. lupus* or *C. dirus* on a morphological or chronological basis (specimens with the typical characters of these two species do not appear in North America until the Illinoian). The earlier material may represent the lineage of *C. armbrusteri* and is listed at this point. The following list is arranged alphabetically by state; specimens examined by me are identified by element, museum number, or both; and selected measurements are found in appendix B (part 13) and appendix C (part 5). Occurrences also are shown on the map in figure 45.

ARIZONA.—Anita, Coconino County; early Irvingtonian (Richard H. Tedford, American Museum of Natural History, pers. comm.); as "*C. nubilus?*" (Hay, 1921:632); mandibular fragment, USNM 10210 A. Of three such specimens from this site, two are referred above to *C. edwardii.* The third mandible is much larger and deeper, and is almost identical in size and proportion with certain specimens from Cumberland Cave.

CALIFORNIA.—Irvington, Alameda County; Irvingtonian; as *C.* cf. *dirus* (Savage, 1951:230). The few pertinent fragments from this site do not seem adequate for identification, but do represent a large canid and may be referable to *C. armbrusteri.*

FLORIDA.—McCleod lime rock mine, 2.5 mi. N Williston, Levy County; Irvingtonian; cranial fragment, AMNH 67286; two maxillary fragments (probably from same individual), AMNH 67287-67288; two mandibular fragments (probably from same individual), AMNH 67289-67290; mandibular fragment, AMNH 67291. The specimens are large, but do not match *C. dirus* in size or other critical characters. Referral to *C. armbrusteri* is supported by the presence of a pronounced buccal cingulum on M1, a prominent deuterocone on P4, a pronounced posteromedial cingulum on p4, and a relatively shallow mandible.

Coleman IIA local fauna, Sumter County; Irvingtonian; as *C. lupus* (Martin, 1974:76); skull without mandibles, UF 11519; cranial fragment, maxillary fragment, and three mandibular fragments, UF 11520; mandibular fragment, two m1, UF 12121; mandibular fragment, UF 11518; two P4, UF 12114; various teeth and postcranial elements, UF. Although Martin (1974:75) considered one skull of *C. armbrusteri* from Cumberland Cave (USNM 7994) to be "essentially identical" to the Coleman skull UF 11519, he assigned these and other specimens from both sites to *C. lupus.* For reasons stated above I recognize *C. armbrusteri* as a distinct species, probably most closely related to *C. rufus,* and the Coleman material seems best referred to *C. armbrusteri.* Some of the specimens are comparatively small, but most dimensions fall within the size range for the Cumberland material. Other characters in which the Coleman wolves resemble *C. armbrusteri,* rather than *C. lupus,* include: the pronounced buccal cingulum, and relatively large, deeply sculptured medial section on M1; the prominent deuterocone on P4; the pronounced posteromedial cingulum extending well behind the third cusp on p4; and the large, well inflated bullae of the one specimen on which they could be evaluated.

MARYLAND.—Cumberland Cave, 4 mi. NW Cumberland, Allegany County; Illinoian (Kurten, 1963:100); as *C. armbrusteri* (Gidley, 1913:98; Gidley and Gazin, 1938:15), as *C. lupus* (Martin, 1974:76); eight skulls, USNM 7994, 8144, 11881, 11883, 11885, 11886, 11887, 12288; 13 mandibular fragments, USNM 7482, 7661, 8144, 8168, 8169, 8172, 11881, 11882, 11887, 11888, 12290, 12293, 12295; P4, USNM 12289; M1, M2, FM P14790.

NEBRASKA.—Rushville fossil quarry, Sheridan County; Yarmouthian (Schultz and Martin, 1970); as *C. dirus nebrascensis* Frick (Schultz and Tanner, 1957:71); maxillary fragment, UN 25691. The features of the M1 in this specimen are not those of *C. dirus*, but fall within the range of variation shown by *C. armbrusteri*.

Angus fossil quarry, Nuckolls County; Yarmouthian; as *C. dirus nebrascensis* Frick (Schultz and Martin, 1970:347). According to Larry D. Martin (Department of Systematics and Ecology, University of Kansas, pers. comm.), this record is based on an ulna, but Merriam (1912:236) reported that in *C. dirus* this element shows no sharp distinguishing characters.

PENNSYLVANIA.—Port Kennedy deposit, Upper Merion Township, Montgomery County; Irvingtonian (probably Yarmouthian: Hibbard, 1958); as possibly *C. indianensis* (Cope, 1899:227). Cope reported three postcranial elements to be larger than those of any wolf known to him. Such material is not reliable in the identification of the dire wolf, and the record herein is listed under *C. armbrusteri*.

SOUTH CAROLINA.—Ashley River, Charleston County; Pleistocene; as *C. occidentalis* (Hay, 1923:365). A mandibular fragment with p4 was compared by Hay to *C. dirus* and *C. lupus*, and was found to be closer in size to the latter. Unfortunately, the single specimen of *C. lupus* used by Hay (USNM 9001) is the largest skull of that species that I have examined, and thus is hardly typical. It seems unlikely that gray wolves of this size ever occurred as far to the southeast as Charleston. But the measurements of depth of jaw (28.0 millimeters) and length of p4 (18.5 millimeters), listed by Hay for the Ashley River specimen, are almost identical to those of several specimens of *C. armbrusteri* from Cumberland Cave and Florida.

TEXAS.—Rock Creek, Briscoe County; Kansan (Hibbard, 1970); as *C. dirus* (Troxell, 1915:633). Troxell referred a tibia and several other postcranial elements to *C. dirus*, solely on the basis of size. According to Stock and Lance (1948), however, the body of *C. dirus* was small relative to its skull. Thus size would not be a reliable character in distinguishing the postcranial skeleton of the dire wolf from that of other large species.

Evolutionary position.—*Canis armbrusteri* is one of several large species that arose from the basal stock of primitive wolves represented by *C. edwardii* and *C. rufus*. Descent could have been directly from either of these latter two species. The presence of specimens of large wolves at several early Irvingtonian sites suggests that divergence between the lineages of *C. armbrusteri* and *C. rufus* occurred early in the Pleistocene, and is evidence for a lengthy independent evolution of the wolf group. Unfortunately, this early material is so fragmentary that it is impossible to determine how many species of wolves are represented. Specimens clearly showing typical characters of *C. lupus* and *C. dirus* do not appear in North America until the Illinoian, at which time the less specialized *C. armbrusteri* was still present. Therefore, it is reasonable to suppose that before then *C. armbrusteri* was the only large wolf in North America, and that it may have occurred over much of the continent. In the Illinoian, *C. armbrusteri* might have become restricted to the east by the initial movement of circumpolar *C. lupus* into the plains and western mountains. *Canis armbrusteri* disappeared by the end of the Illinoian, but we do not know if its lineage ended then or if it gave rise to *C. dirus*, as suggested by Martin (1974:76). This latter hypothesis is not adequately supported by available morphological evidence, and perhaps *C. dirus* was a replacement for, rather than a descendent of, *C. armbrusteri*.

Canis lupus Linnaeus

1758. *Canis lupus* Linnaeus, Systema Nautrae, 10th ed., p. 39. Type locality, Sweden.

1910. *Canis occidentalis furlongi* Merriam, Univ. California Publ. Bull. Dept. Geol., 5:393. Type from Rancho La Brea, Los Angeles County, California. Valid as a subspecies of *C. lupus*.

1912. *Canis milleri* Merriam, Mem. Univ. California, 1:247. Type from Rancho

La Brea, Los Angeles County, California. A synonym of *C. lupus furlongi.*
1918. *Aenocyon milleri,* Merriam, Univ. California Publ. Bull. Dept. Geol., 10:533.

In addition to those listed above, 24 names based on North American Recent specimens are available for use at the subspecific level. These are to be found in the systematic revision by Goldman (1944), and also are listed by Hall and Kelson (1959:847-851).

Type.—None designated.

Geological distribution.—Late Irvingtonian to Recent in North America.

Geographical distribution.—Pleistocene and early Recent records from Alberta, Saskatchewan, Yukon, Alaska, Arizona, Arkansas, California, Colorado, Georgia, Idaho, Illinois, Kansas, Michigan, Minnesota, Nebraska, Nevada, New Mexico, Oklahoma, Oregon, Pennsylvania, Texas, Virginia, Wisconsin, Wyoming, Nuevo Leon, and many localities in Eurasia. Historical range throughout Eurasia, except tropical forests of southeastern corner; throughout North America, except parts of southeastern quarter, southern and coastal Mexico, Central America, Baja California, and most of California; and on most adjacent continental islands. Presently extirpated in many areas settled by man, including most of Europe and the 48 southern continental states of the United States.

Description.—Size large for the genus; skull usually large with mostly broad proportions; rostrum elongated, usually relatively broad and deep; braincase relatively small, not much inflated dorsoposteriorly; postorbital constriction elongated, narrow lateromedially; zygomata thick, deep, broadly flaring; orbits relatively small; frontals usually well elevated above rostrum, prominently convex, forming relatively broad shield; temporal ridges sharp, often obscuring frontal suture, usually joining anterior to coronal suture; sagittal crest prominent, sharp dorsally; supraoccipital shield large; external side of occipital well ossified; tympanic bullae usually moderate in size, not much inflated; mandible thick and deep, ventral margin not convex when viewed from side, toothrow bowed outward in center; incisors relatively large; upper canines prominent, thick anteroposteriorly, alveoli set relatively high in premaxillae, ventral tips usually not extending to level of anterior mental foramina when jaws are closed; premolars relatively broad; P4 usually lacking prominent deuterocone and lingual cingulum; M1 having relatively large paracone and metacone, relatively small medial section without trenchant cusps, the metaconule reduced and not well separated from protocone; M1 lacking pronounced buccal cingulum; M2 usually relatively small; p2 often with posterior cusp; p3 usually with second and third cusp; p4 with second cusp, sometimes lacking third cusp, usually without posteromedial cingulum extending behind third cusp; m1 relatively broad, usually having relatively small talonid. For details on pelage and postcranial skeleton see Goldman (1944), Iljin (1941), Mech (1970), and Hildebrand (1952a, 1952b, 1954).

Comparison with C. latrans.—Usually much larger; skull larger and relatively broader in most dimensions; rostrum relatively broader and deeper, especially in posterior half, flaring out more anterolaterally; braincase relatively smaller, less inflated dorsally, never broader at level of parietotemporal sutures than at base; postorbital constriction narrower, more elongated, rising more steeply into frontal region; zygomata deeper, thicker, more broadly flaring; orbits relatively smaller; frontals more elevated above rostrum, more depressed medially, more prominently convex, forming broader shield; temporal ridges sharper, more often obscuring frontal suture, and joining anterior, rather than posterior, to coronal suture; sagittal crest more prominent; supraoccipital shield broader, projecting farther posteriorly; external side of occipital more ossified, seldom with any trace of thin-walled projection dorsal to foramen magnum; occipital condyles extending farther transversely; tympanic bullae usually less inflated; mandible thicker and deeper, ventral edge less convex when viewed from side, toothrow more bowed outward in center; incisors larger, extending farther transversely; upper canines thicker anteroposteriorly, their alveoli set more dorsally in premaxillae, not extending so far ventrally; premolars broader with less trenchant cusps, usually more closely set in jaws; P4 with deuterocone and lingual cingulum much less prominent or absent; M1 having relatively larger paracone and metacone, relatively smaller medial section with less trenchant cusps, hypocone less prominent, metaconule smaller and less distinct from protocone, protoconule often less distinct from protocone, anterior cingulum less pronounced, buccal cingulum less pronounced or absent; M2 usually relatively (and occasionally absolutely) smaller with less trenchant cusps; p2 more often with posterior cusp; p4 with posterior cusps more reduced, more often lacking well developed third cusp and posteromedial cingulum; m1 relatively broader, metaconid less prominent and not projecting so far medially, heel relatively smaller; m2 and talonid of m1 with less trenchant cusps. Hildebrand (1952a, 1954) discussed differences between the postcranial skeletons of *C. lupus* and *C. latrans*; Atkins and Dillon (1971) listed distinguishing features of the cerebellum.

Other comparisons.—See accounts of *C. rufus*, *C. armbrusteri*, *C. familiaris*, and *C. dirus*.

Remarks.—The gray wolf is probably the most widely distributed and most naturally successful species of *Canis* ever to exist. Its size, intelligence, and social nature are singularly adapted for its role as the major predator of northern ungulates. The systematics of the species have long been a source of confusion, and are still not completely understood. Especially difficult problems involve the status of the small Recent subspecies or species of wolves that existed all along the southern margins of the range of *C. lupus*.

Imaizumi (1970a, 1970b) recently raised the extinct Japanese wolf, *hodophilax*, back to the level of a full species. The wolf of China and central Asia (*chanco*), and of India and the Near East (*pallipes*) are probably not more than subspecifically distinct from *C. lupus*, but adequate series of specimens from these vast regions never have been studied in detail. On the basis of cranial measurements provided by Pocock (1935:671), and the few specimens that I have seen, *pallipes* seems to be a highly variable entity with a cranial size range bridging the gap between North American *C. lupus* and *C. latrans*. Even more interesting in this regard is *arabs* of southern Arabia, of which the cranial measurements listed by Harrison (1968:203) indicate an animal averaging not much larger than *C. latrans*. Lawrence (1966:57) suggested that *arabs* may have been influenced by hybridization with *C. familiaris*. Harrison (1973:190), however, reported that all available skulls of *arabs* could be distinguished by the relatively greater size and inflation of their bullae. Two other southern subspecies listed by Ellerman and Morrison-Scott (1951:218-220), *C. lupus italicus* of Italy, and *C. l. signatus* of Spain, had been synonymized under *C. l. lupus* of most of Eurasia by Pocock (1935). One more named subspecies, *C. l. deitanus*, was based only on two live animals from southeastern Spain. Miller (1912c:315) noted that they had a "general appearance much as in *C. aureus*." On the basis of this description, Pocock (1935:653) suggested the possibility that *deitanus* was a representative of the North African jackal. The question apparently never has been resolved.

For North America no attempt has been made to go beyond previous studies in assessing the intraspecific relationships within *C. lupus*. All of the names and their areas of application, summarized by Hall and Kelson (1959:847-851), are maintained in this paper. Some comment, however, is necessary regarding a confusing situation on the Arctic islands. Anderson (1943) and Goldman (1944) considered that four subspecies inhabited this region (see map, Fig. 2). Manning and Macpherson (1958), following extensive statistical analysis, concluded that the kind of wolf represented by a series of eight skulls (including only two adults) collected in 1914-1916 on Banks Island, and described as *C. lupus bernardi* by Anderson (1943), had been replaced by a different kind of wolf, represented by 16 specimens collected in 1953-1955, that seemed closest to *C. l. arctos* of Prince Patrick and Ellesmere islands. I did not measure the series of specimens taken on Banks Island in 1914-1916, but skulls of six males and two females collected there in 1953-1955 were suitable for inclusion in multivariate analyses. The series of males demonstrates a consistently high statistical distance from each subspecies of gray wolf, including *arctos*. The most striking character of the recently collected Banks Island skulls is their great maximum width across the upper cheek teeth. In all but one of these specimens this width actually exceeds the alveolar length from P1 to M2. In most other skulls of *C. lupus*, including all but one of the 21 *arctos* that I measured, the length was greater than the width. The Banks Island skulls also differ from *arctos*, and most other subspecies of *C. lupus*, in their greater width of frontal shield. Although I agree with Manning and Macpherson (1958:43) that the more recently collected skulls from Banks Island differ from Anderson's description of *C. l. bernardi*, I am not so certain that these specimens may be "assigned to *C. l. arctos* with confidence."

The most critical problem that concerned the Recent wolves of the New World was the relationship of *C. lupus* with *C. rufus* of the southeast. Information provided in the previous part of this paper, and in the account of *C. rufus* in this part, has to me confirmed the specific status of the red wolf. Nonetheless, the paucity of available material from the eastern United States gives an incomplete picture of the original situation in that region. In the above account of *C. rufus*, I discuss a number of eastern specimens which seem best referred to that species. Various other fragments from the east, including many from archeological sites and not listed by Goldman (1944) or Hall and Kelson (1959), probably represent *C. lupus lycaon*. Such specimens examined by me (indicated by element), or reported by others, include the following.

Tick Creek Cave site, 12 mi. W Rolla, Phelps County, Missouri; as *C. lupus* (Parmalee, 1965:19). Parmalee also reported that some remains from this site may represent *C. rufus*.

Bell site, 5 mi. W Oshgosh, Winnebago County, Wisconsin; as *C. lupus* (Parmalee, 1963:61).

Raddatz rock shelter, central Sauk County, Wisconsin; as *C. lupus* (Parmalee, 1959b:85); maxillary fragment, mandibular fragment, ISM.

Moccasin Bluff site, west of Buchanan, Berrien County, Michigan; as "wolf" (Cleland, 1966:205).

Anker site, Cook County, Illinois; as *C. lupus* (Parmalee, 1959a:91); two maxillary fragments, premaxillary fragment, two mandibular fragments (probably all from same individual), ISM.

Fisher site, south bank Des Plaines River, Will County, Illinois; as *C. lupus* (Parmalee, 1962b:402); mandibular fragment, ISM.

Kingston Lake site, 15 mi. SW Peoria, Peoria County, Illinois; as *C. lupus* (Parmalee, 1962a:10).

Hummel Camp site, 1 mi. S London Mills, Fulton County, Illinois; as *C. nubilus* (Cole and Deuel, 1937:265).

Weaver site, Fulton County, Illinois; as *C. lupus* (Parmalee, 1959a:91).

Clear Lake site, Tazewell County, Illinois; as *C. lupus* (Parmalee, 1959a:91); maxillary fragment, mandibular fragment, ISM.

Busch Estate site, Pike County, Illinois; as *C. lupus* (Parmalee, 1959a:91).

Knight site, Calhoun County, Illinois; as *C. lupus* (Parmalee, 1959a:91).

Snyders site, Calhoun County, Illinois; as *C. lupus* (Parmalee, 1959a:91).

Apple Creek site, Greene County, Illinois; mandibular fragment, ISM.

Cahokia site, near East St. Louis, Madison County, Illinois; as *C. lupus* (Parmalee, 1957:239).

Palestine site, Palestine, Crawford County, Illinois; as *C. lupus* (Parmalee and Stephens, 1972:71); mandibular fragment, ISM.

Sugar Camp Hill site, Williamson County, Illinois; as *C. lupus* (Parmalee, 1959a:91).

Fifield site, Porter County, Indiana; as "gray wolf? *Canis lupus*" (Parmalee, 1972:205).

Breck Smith Cave, 8 mi. W Lexington, Fayette County, Kentucky; as "wolf" (Miller, 1922).

Citico Mound, near Citico Creek, Hamilton County, Tennessee; as *C. rufus floridanus* (Kellogg,

1939:267), as *C. lupus lycaon* (Goldman, 1944: 441); mandibular fragment, USNM.

Madisonville ancient cemetery, Cincinnati vicinity, Hamilton County, Ohio; as *C. lupus* (Langdon, 1881:299).

Hobson site, near Middleport, Meigs County, Ohio; as *C. lupus* (Murphy, 1968:12).

Fairchance Mound, Moundsville, Marshall County, West Virginia; as *C.* cf. *lupus* (Guilday and Tanner, 1966:42).

Mount Carbon site, 3.5 mi. SW Montgomery, Fayette County, West Virginia; as *C. lupus* (Guilday and Tanner, 1965:2); m1, CM.

Doepkin's Farm site, U.S. Hwy. 50, Anne Arundel County, Maryland; maxillary fragment, Doepkin's Farm collection.

Quaker State Rockshelter, 3 mi. SE Franklin, Venango County, Pennsylvania; as *C. lupus* (Guilday and Tanner, 1962:134).

Sheep Rock shelter, west bank Raystown branch Juanita River, Huntingdon County, Pennsylvania; as *C. lupus* (Guilday and Parmalee, 1965:38).

Johnston site, Indiana County, Pennsylvania; maxillary fragment, CM.

Hartley site, Greene County, Pennsylvania; mandibular fragment, CM.

Eschelman site, 3 mi. S Columbia, Lancaster County, Pennsylvania; as *C. lupus* (Guilday, Parmalee, and Tanner, 1962:64); three mandibular fragments, CM. The red wolf, *C. rufus*, is also represented by material from this site.

Lewiston Mound, Lewiston, Niagara County, New York; as "wolf" (Ritchie, 1969:218).

Garoga site, Fulton County, New York; maxillary fragment, two mandibular fragments, CM.

Frontenac Island, Cayuga Lake, Cayuga County, New York; as *C. lupus* (Ritchie, 1969:106); maxillary fragment, CM.

Lamoka Lake site, near Tyrone, Schuyler County, New York; as *C. lupus* (Guilday, 1969:55).

Sawyer's Island, near Boothbay, Lincoln County, Maine; as *C. occidentalis* (Loomis and Young, 1912:27).

In addition to the above records, Manville and Sturtevant (1966) reported the presence of two Indian artifacts, containing parts of wolf skulls, in the collection of the Skokloster Castle Museum in Sweden. The specimens had probably been obtained from Indians near the Swedish colony on the Delaware River, or the Dutch colony on the Hudson River. The material was identified as *C. lupus lycaon*, and the measurements provided indicate that the gray wolf, rather than the red wolf, is represented.

Whereas there is sometimes difficulty in distinguishing specimens of *C. lupus* and *C. rufus*, cranial material of *C. lupus* and *C. latrans* can always be separated. The clear distinction of the two species was recognized by American taxonomists at least as early as Audubon and Bachman (1851). Baird (1857:104) adequately described some of the major cranial differences between the gray wolf and coyote. Cope (1879:184) was the first to point out the discriminating features of the cusps on the medial section of M1. Gidley (1913:98-102) listed what he considered to be diagnostic characters of the lower dentition, but, as explained by Jackson (1951:242), these characters are not always reliable. The most thorough discussion of the differences in proportions and other characters, between the skulls of *C. lupus* and *C. latrans*, was that provided by Lawrence and Bossert (1967).

The coyote and gray wolf shared a large part of their respective ranges in North America, but hybridization under completely natural conditions occurred rarely, if ever. Interbreeding in eastern Canada, caused largely by recent human environmental disruption, has resulted in the production of some specimens with intermediate characters.

Fossils of wolves, other than *C. dirus*, are comparatively rare in North America, and it sometimes is difficult to determine what species are represented. Martin (1974:76) considered *C. armbrusteri* of Maryland and Florida to be synonymous with *C. lupus*, but, as explained previously, the two are distinct. One named Pleistocene species that now can be synonymized with a subspecies of *C. lupus* is *C. (Aenocyon) milleri* Merriam from Rancho La Brea. The single specimen on which the species was based falls within the morphological range of *C. lupus*, and seems best referred to *C. lupus furlongi* (see account of Rancho La Brea, below).

Fossil record.—Most fossil wolf material is so fragmentary that determination as to species is difficult. The first specimens that show the specific characters of *C. lupus* appear in Illinoian deposits. Several pre-Illinoian fragments that had been referred by

Fig. 51.—Map showing localities (black dots) of fossil *C. lupus*. Because of the scale of the map, it was not possible to plot all localities in crowded areas.

others to this species, are discussed above in the accounts of *C. edwardii*, *C. rufus*, and *C. armbrusteri*. The following list contains additional literary references to fossil *C. lupus*. The list is arranged alphabetically by state and province, and geographically (north to south, west to east) within states and provinces, except that Canadian provinces are listed first and Mexican states last. Specimens examined by me are identified by element, museum number, or both; and selected measurements are found in appendix B (part 14) and appendix C (part 6). Occurrences also are shown on the map in figure 51.

ALBERTA.—Medicine Hat (north of); early Recent; as *C. lupus* (Churcher, 1969b:181).

Medicine Hat; Sangamon; as *C. lupus* (Churcher, 1970:62).

Island Bluff, near Medicine Hat; Sangamon; as *C. lupus* (Churcher, 1969b:181).

Mitchell Bluff, near Medicine Hat; late Pleistocene; as *C. lupus* (Churcher, 1969a:2).

SASKATCHEWAN.—Fort Qu'Apelle; late Pleistocene; tibia, CNM 12178 (see also Khan, 1970:13).

YUKON.—Old Crow area; late Pleistocene; as "wolf" (Geist, 1955:1702); mandibular fragment, CNM 17311.

Hunker Creek vicinity, Klondike River; late Pleistocene; as "*Canis* (wolf)" (Quackenbush, 1909:127); skull without mandibles, CNM 9929. The specimen shares certain characters with Recent wolves of the Arctic islands, including a broad rostrum, crowded toothrow, and relatively large carnassials.

Gold Run Creek, 30 mi. SE Dawson; Wisconsin (C-14 dates: 22,200 and 32,250 B.P.); as *C.* cf. *lupus* (Harington and Clulow, 1973:699).

Quartz Creek; late Pleistocene; cranial fragment, CNM 17311.

ALASKA.—Historic Bluff, entrance to Eschscholtz Bay; late Pleistocene; as "*Canis* (wolf)" (Quackenbush, 1909:97); maxillary fragment and other cranial elements, AMNH 13753.

Buckland River, southeast of Eschscholtz Bay; late Pleistocene; as "*Canis* (wolf)" (Quackenbush, 1909:120).

Fairbanks (near); Illinoian; as "*Canis* sp. (wolf)" (Pewe and Hopkins, 1967:267).

Fairbanks Creek Mine, near Fairbanks; Wisconsin; as *C. lupus* (Guthrie, 1968:352).

Engineer Creek Mine, near Fairbanks; Wisconsin; as *C. lupus* (Guthrie, 1968:352).

Gold Hill Mine, near Fairbanks; Wisconsin; as *C. lupus* (Guthrie, 1968:352).

Cripple Creek Mine, near Fairbanks; Wisconsin; as *C. lupus* (Guthrie, 1968:352).

ARIZONA.—Ventana Cave, Papago Indian Reservation, Pima County; Wisconsin (Hibbard, 1958); as *C. lupus* (Colbert, 1950:132).

Papago Springs Cave, southeast of Sonoita, Santa Cruz County; Wisconsin (Hibbard, 1958); as *C. nubilus* (Skinner, 1942:164).

ARKANSAS.—Conard fissure, 15 mi. S Harrison, Newton County; Illinoian (Kurten, 1963:100); as "*C. occidentalis?*" (Brown, 1908:182); cranial fragment, isolated teeth, postcranial fragments, AMNH 11762; mandibular fragment, AMNH 11761.

CALIFORNIA.—Samwel Cave, Shasta Lake, Shasta County; Wisconsin (Hibbard, 1958); "the specimen appears to resemble the northern wolves; for example *C. l. pambasileus*, rather than specimens from the southern part of the range of *C. lupus*" (Graham, 1959:58).

Potter Creek Cave, 1 mi. SE Baird, Shasta County; Wisconsin (Hibbard, 1958); as *C. lupus* (Kurten and Anderson, 1972:37); mandibular fragment, UCMP 5018.

McKittrick tar seeps, Kern County; Wisconsin; as "gray or timber wolf" (Sternberg, 1928:226), as *Aenocyon* near *milleri* (Schultz, 1938b:169). Sternberg reported that he had collected ten specimens of gray wolves at McKittrick, but Schultz did not mention the presence of *C. lupus* or *C. furlongi* at the site. Schultz did note that two M1 seemed best referred to *Aenocyon milleri*, which in this paper is considered a synonym of *C. lupus furlongi*.

Maricopa Brea, near Maricopa, Kern County; Wisconsin; skull and mandible, LACM 18419; skull without mandibles, LACM 21921; maxillary and mandibular fragments, LACM 18798; three maxillary fragments, LACM 20531 and two unnumbered; cranial fragment, LACM; two mandibular fragments, LACM 17890, 22288; M1, LACM. These specimens were identified by me from among a larger number of specimens of *C. dirus* in the hitherto unreported Maricopa Brea collection.

Rancho La Brea, Los Angeles, Los Angeles County; Wisconsin; as *C. occidentalis furlongi* (Merriam, 1910:393; 1912:251; Schultz, 1938b:163; Goldman, 1944:399; Anderson, 1968:26), as *C. furlongi* (Hay, 1927:184; Stock, 1956:33), as *C. lupus* (Hibbard, 1958:18); eight skulls without mandibles, LACM 2300-44, 2300-56, 2300-353, 2300-384, 2600-1, 2600-5, 236(315), one unnumbered; cranial fragment, LACM; two mandibles, LACM 2301-L476, 2301-L495; incomplete skull, UCMP 19792; maxillary fragments, UCMP 10733; maxillary fragment, mandibular fragment, UCMP 11283. Merriam (1910) considered that certain specimens from Rancho La Brea represented an animal closely related to the modern gray wolf. His name for this animal, *C. occidentalis furlongi*, indicated his recognition of it as a subspecies of the North American gray wolf which was then (1910) often referred to as *C. occidentalis*. Miller (1912b), however, restricted the name *occidentalis* to the interior forests of northern Canada, and Hay (1927:184) considered it improbable that a subspecies of *occidentalis* ever would have been present in southern California. Subsequently, various authors either followed Hay in listing *furlongi* as a full species, or continued to use the trinomial *C. occidentalis furlongi*. Since all of the Recent gray wolves of North America, including *occidentalis*, were arranged as subspecies of *C. lupus* by Goldman (1944), and since Merriam's original intention was obviously to recognize *furlongi* as a subspecies of gray wolf, the proper name for the animal in question would be *C. lupus furlongi*. Merriam (1910, 1912) based his descriptions of *furlongi* on three fragmentary specimens in the University of California Museum. The material was said to be smaller than that of the much more abundant *C. dirus* from Rancho La Brea, and to be characterized by a more prominent hypocone on M1. Subsequently, Stock (1929:20) reported that 20 individuals of *furlongi* from Rancho La Brea were represented in the collections of the Los Angeles County Museum, but later (1956:33) he wrote that "only eight specimens have been recognized in the Museum collections." Considering that taxonomists of the early twentieth century sometimes named species on the basis of less critical analysis than is usual today, it might be tempting to write off the few specimens of *furlongi* as small, aberrant examples of *C. dirus* (of which 1,646 individuals were reported to be represented in the LACM collections from Rancho La Brea by Marcus, 1960:5). There is no question, however, that *C. lupus* is also present. The specimens of *C. dirus* from Rancho La Brea are remarkably consistent in certain critical characters, and while going through the unlabeled collection of wolves at the Los Angeles County Museum, I found the eight above listed skulls of *C. lupus* to stand out clearly from the others. I think that Stock (1956: 33) must have been referring to the same eight specimens. Each of these skulls is from an adult animal, and has each of the following characters for which it can be evaluated: frontals depressed medially; temporal ridges sharp; orbital angle under 47°; supraoccipital shield broad, not projecting far posteriorly; postpalatine foramina placed well anterior to posterior edges of P4; optic foramen and anterior lacerated foramen well separated; vertical plates of palatines not broadly flaring anteriorly; posterior end of vomer extending well behind posterior nasal opening; M1 with large hypocone, its ridge extending completely or almost completely around anterior base of protocone. These characters, along with over-all moderate size, distinguish the LACM specimens, as well as those in the UCMP, from *C. dirus*. In addition to these skulls, I found a cranial fragment having the characters of *C. lupus*, and two mandibles characterized by small size, a prominent posterior cusp on p2, no posterior cingulum on p4, and a relatively high set heel on m1. Most of the specimens of *furlongi* have comparatively large teeth, especially carnassials, a broad rostrum, and a relatively broad frontal shield. These characters are shared by *C. dirus*, and might suggest that *furlongi* represents an evolutionary transition between *C. dirus* and *C. lupus*. But there is no chronological evidence to support this view; material of both species was found together in the same pits at Rancho La Brea. Furthermore, in the great majority of characters, the material from this site shows no tendency toward blending; each specimen can be unquestionably referred to either *C. lupus* or *C. dirus*. Large carnassials, and a broad rostrum and frontal shield are also present in specimens of modern gray wolves of the Arctic islands (see "remarks," above). Macpherson (1965:164) hypothesized that wolves with such characters had been isolated by late Pleistocene glaciation in a Pearyland refugium, and had subsequently spread back across the Arctic. Possibly this population had once occupied a large northern area, and had then been driven by the Wisconsin glaciation both northeastward into Pearyland, and southwestward as far as Rancho La Brea. Not all specimens of *C. lupus* from this site are as massive as Recent Arctic wolves, and, as noted by Merriam (1912:253) there is considerable variation in size of teeth.

Rancho La Brea, Los Angeles, Los Angeles County; Wisconsin; as *C. milleri* (Merriam, 1912:24), as *Aenocyon milleri* (Merriam, 1918:533), as *C. lupus* (Martin, 1974:76); skull with mandibles, UCMP 11257. According to Merriam's original description, the single known specimen from Rancho La Brea is intermediate in characters between *C. lupus* and *C. dirus*. He compared the specimen most critically with *C. dirus*, from which it was said to differ in having lesser size, a smaller frontal shield, lower sagittal crest, less overhang of inion, more anteriorly placed postpalatine foramina, and more prominent hypocone on M1. Merriam reported *milleri* to have a much broader palate and much more massive dentition than *C. lupus*. Later (1918) he observed that the characters of *milleri* justified placing it together with *dirus* in the new genus *Aenocyon*. Subsequently, according to Stock, Lance, and Nigra (1946:109), the validity of *milleri* as a species was questioned, but they did not indicate whether referral to *C. dirus* or *C. lupus* was being considered. Martin (1974:76) recognized *C. milleri* as a synonym of *C. lupus*. Had Merriam been able to examine the eight skulls of *C. lupus furlongi* from Rancho La Brea in the Los Angeles County Museum, and a series of Recent *C. lupus* from the Arctic islands, he might not have established *milleri* as a separate species. As he himself observed (1912:247), the combination of characters found in *milleri* is approached most closely in *C. l. furlongi*, and several of the LACM specimens are almost identical to *milleri*. In its relatively large carnassials and unusually broad rostrum, *milleri* resembles some of the living wolves of the Arctic. Merriam's (1918) association of the specimen with the genus *Aenocyon* was a mistake, as the specimen differs from the dire wolf in all of the critical characters mentioned above in the discussion of *C. lupus furlongi*.

Schuiling Cave, 2 mi. SE Newberry, San Bernardino County; late Pleistocene; as *C.* cf. *lupus* (Downs, et al., 1959:9).

COLORADO.—Chimney Rock animal trap, Larimer County; late Pleistocene or early Recent; as *C. lupus* (Hager, 1972:65).

GEORGIA.—Ladds, near Cartersville, Bartow County; late Pleistocene; as *C.* cf. *lupus* (Ray, 1967: 133); M1, USNM 23698.

IDAHO.—Jaguar Cave, Beaverhead Mountains, Lemhi County; late Wisconsin (C-14 dates: 10,370 ±350 and 11,580±250 B.P.); as *C. lupus* (Kurten and Anderson, 1972:24).

Moonshiner Cave, Bingham County; late Wisconsin or early Recent; as *C. lupus* (Kurten and Anderson, 1972:37).

ILLINOIS.—Polecat Creek gravel pits, 1 mi. S Ashmore, Coles County; late Wisconsin (Hibbard, et al., 1965); as *C. lupus* (Galbreath, 1938:309).

KANSAS.—Goodland, Sherman County; late Pleistocene or early Recent; as *C. lupus* (Williston, 1898:93; Hibbard, Frye, and Leonard, 1944:10); skull with mandibles, KU 2851. The specimen resembles skulls of *C. lupus baileyi* from the southwest, and is smaller than skulls of *C. l. nubilus* collected on the Great Plains in historical time.

MICHIGAN.—Millington, Tuscola County; late Wisconsin; as *C. lupus* (Wilson, 1967:211); pair of mandibles, upper incisors, UMMP 33770.

MINNESOTA.—Itasca bison site, Clearwater County; late Pleistocene; as "wolf" (Shay, 1963:48).

MISSOURI.—Brynjulfson Caves, 6 mi. SSE Columbia, Boone County; late Wisconsin (about 10,000 B.P.); as *C. lupus* (Parmalee and Oesch, 1972:29).

NEBRASKA.—Hay Springs quarry, Sheridan County; probably Illinoian (Hibbard, 1958); as *C.* cf. *occidentalis* (Matthew, 1918; Schultz, 1934:369); rostral fragment, AMNH (Frick Collection) 25511; mandibular fragment, UN 2912.

Mullen, Cherry County; late Irvingtonian (Kurten, 1974:7); as *Aenocyon dirus* (Martin, 1972:174); maxillary fragment, UN 39337; mandibular fragment, UN 26117. In size and other characters, these specimens resemble *C. lupus,* not *C. dirus.*

Freedom (near), Frontier County; late Pleistocene; mandible, UN 2911.

Republican River, 1 mi. S Guide Rock, Webster County; late Pleistocene or early Recent; cranial fragments, USNM 18749.

NEVADA.—Lake Lahontan, near Fallon, Churchill County; early Recent; as *C. lupus* (Morrison, 1964:73).

Smith Creek Cave, Baker, White Pine County; late Pleistocene; cranial fragment, LACM 7190.

NEW MEXICO.—Isleta Caves, 8 mi. W Isleta, Bernalillo County; late Wisconsin; as *C.* cf. *lupus* (Harris and Findley, 1964:115), as *C. lupus* (Anderson, 1968:22).

Blackwater Draw, near Clovis, Curry County; Wisconsin (Lundelius, 1967:301); crushed skull, TM 937-521; mandible, TM 937-895; two M1, TM 937-885, 937-905.

Burnet Cave, 50 mi. W Carlsbad, Eddy County; Wisconsin (Hibbard, 1958); as *C. nubilus* (Schultz and Howard, 1935:284); mandible, UN 14004.

Hermit's Cave, east slope of Guadalupe Mountains, Eddy County; late Wisconsin (C-14 dates: 11,850±350 and 12,900±350 B.P.; Schultz, Martin, and Tanner, 1970:119); two maxillary fragments and two mandibular fragments (probably all from same individual), UN 19211, 19217, 19218, 19220; pair of mandibles, UN 19216. These specimens were associated with a man-made hearth from which the C-14 dates were obtained. At least one of the fragments is charred, as from fire, and the single skull, apparently represented by the first four fragments listed above, may have been deliberately broken apart. Schultz, Martin, and Tanner (1970) recorded only *C. dirus* from Hermit's Cave, and that species is present, but the five specimens listed above unquestionably represent *C. lupus.*

Dark Canyon Cave, Eddy County; late Pleistocene; mandible, LACM 1644.

OKLAHOMA.—Selman Cave system, 7 mi. SW Freedom, Woodward County; Recent, as *C. lupus* (Black and Best, 1972); mandible, collection of Troy L. Best.

Afton, Ottawa County; Wisconsin (Kurten, 1974:9); as *C. nubilus* (Hay, 1920:129); skull with mandibles, USNM 196943; three mandibles, USNM 196946, 196947, 196948; P4, C1, USNM 9128; eight canine teeth, USNM 9129; premaxillary fragment, P4, M1, p4, USNM 9130.

Arkansas River, Le Flore County; Pleistocene; cranial fragment, AMNH 32669.

OREGON.—Bend (near), Deschutes County; late Pleistocene; tibia, CNM 12178.

Fossil Lake, Lake County; early or middle Wisconsin (Allison, 1966:32); as *C.* cf. *occidentalis* (Elftman, 1931:7).

PENNSYLVANIA.—Crystal Hill Cave, 3 mi. W Stroudsburg, Monroe County; late Pleistocene or early Recent; as *C. lupus* (Leidy, 1889; Hay, 1923:310).

TEXAS.—Lubbock Reservoir, Lubbock County; Wisconsin (Lundelius, 1967:302); mandible, TM 892-255.

Schulze Cave, 28 mi. NE Rock Springs, Edwards County; Wisconsin or early Recent; as *C.* cf. *lupus* (Dalquest, Roth, and Judd, 1969:256).

VIRGINIA.—Natural Chimneys, 1 mi. N Mt. Solon, Augusta County; late Wisconsin (*ca.* 10,000-15,000 B.P.); as *C.* cf. *lupus* (Guilday, 1962:94).

WISCONSIN.—Blue Mounds, Dane County; late Pleistocene; as *C. occidentalis* (Hay, 1918:347; 1923:341).

WYOMING.—Little Box Elder Cave, west of Douglas, Converse County; Wisconsin; as *C. lupus* (Anderson, 1968:25); two mandibles from subadult individual, UColo 22287, 24683.

Bell Cave, Albany County; Wisconsin to early Recent; as *C. lupus* (Anderson, 1974:81).

NUEVO LEON.—San Josecito Cave, near Aramberri; Wisconsin; skull without mandibles, LACM 192-3017; mandibular fragment, LACM 192-28338. In the large collection of canid material from San Josecito Cave, the gray wolf is represented only by these two specimens which probably belonged to the same individual. The skull is much smaller than those of *C. dirus* from the same site, and, indeed, is the smallest skull of an adult *C. lupus* that I have examined. I am indebted to Richard L. Reynolds of the Los Angeles County Museum for recognizing the presence of these specimens in the collection from San Josecito Cave, and for loaning them to me.

Evolutionary position.—The North American gray wolf, like many of our other larger mammals, appears to be a comparatively late immigrant from the Old World. Kurten

(1968:109-110) traced the probable evolution of the species in Europe from a relatively small ancestor. The primitive stock that gave rise to *C. lupus* in the Old World was in all likelihood the same that is represented in North America by *C. edwardii* and early specimens of *C. rufus*. At various times in the Pleistocene, factors associated with glaciation presumably divided this widespread basic stock, and permitted development of several species. While *C. rufus*, *C. armbrusteri*, and *C. dirus* evolved in the New World, *C. lupus* arose in Eurasia and apparently became the species of wolf most suited for the challenging environment of the late Quaternary. We do not know at what point the gray wolf completed its occupation of Eurasia and began to move across the Bering Strait, but conceivably this could have been as early as the Kansan glaciation. A few pre-Illinoian fragments had been questionably assigned to *C. lupus* by previous authors, but at present these specimens seem best referred to *C. rufus* or *C. armbrusteri*. The earliest material that clearly displays the specific characters of *C. lupus* is that from the Illinoian deposits at Hay Springs and Mullen in Nebraska. Illinoian specimens also have been reported from the Conard fissure, Arkansas, and from near Fairbanks, Alaska. It can be reasonably assumed that the species was able to cross the Bering Land Bridge in the Illinoian, and that it eventually established itself in some parts of North America. Glacial movement may have been responsible for the initial appearance of *C. lupus* in the central United States.

The only known Sangamon records of the species in North America are based on a few fragments collected near Medicine Hat, Alberta (Churcher, 1969b:181; 1970:62). Possibly the cold-adapted *C. lupus* had withdrawn from more southerly regions during that interglacial period. The number and distribution of Wisconsin records is much greater, but the fossil history of the species is comparatively poor, and few occurrences are represented by well preserved cranial material.

So fragmentary the material, so variable the existing gray wolf, and so incomplete our understanding of Pleistocene chronology, that it is difficult to assess the factors leading to the present situation. A number of the specimens discussed in the above list do not appear to differ significantly from specimens of *C. lupus* taken in the same areas in historic time. The most interesting Wisconsin specimens are the massive skulls from Rancho La Brea and the Yukon, that resemble the skulls of modern Arctic wolves. A population of wolves with such skulls may have been widespread in the north at the beginning of the Wisconsin, or, more likely, at the beginning of the last major stadial. The glacial movement may then have split the population, one element moving into the Pearyland refugium where it survived, and the other element being driven southward where it either became extinct or was eventually absorbed into other populations of *C. lupus*. A few of the specimens from Rancho La Brea are small, and there are also unusually small skulls of *C. lupus* from Goodland, Kansas and San Josecito Cave, Nuevo Leon. These specimens may represent the result of character displacement following an initial late Pleistocene or early Recent movement of *C. lupus* into areas where *C. dirus* still predominated. Eventually, *C. lupus* prevailed over the dire wolf, either through competition or because of external factors, and established itself as the major large predator of most of North America.

Canis familiaris Linnaeus

1758. *Canis familiaris* Linnaeus, Systema Naturae, 10th ed., p. 38.
1938. *Canis petrolei* Stock, Bull. S. California Acad. Sci., 37:50. Type from Rancho La Brea, Los Angeles County, California. The type and only known specimen, originally described by Stock as "a coyote-like wolf jaw," probably rep-

resents *C. familiaris*. A study on this and other specimens of early domestic dogs from Rancho La Brea is being made by Richard L. Reynolds of the Los Angeles County Museum.

Type.—None designated.

Geological distribution.—Late Rancholabrean to Recent.

Geographical distribution.—World-wide in association with man.

Description.—Exceptionally variable in size and other characters because of the influence of domestication. Less specialized breeds characterized as follows: moderate size; skull medium-sized, relatively broad in most proportions; rostrum usually relatively short, broad, and deep; braincase relatively small, not much inflated dorosposteriorly, broadly based, set low relative to other parts of skull; postorbital constriction usually elongated, broad lateromedially, rising very steeply into frontal region; zygomata relatively thick, deep, moderately flaring; orbits relatively small; frontals rising steeply above rostrum, prominently convex, forming broad shield with bulging postorbital processes; temporal ridges usually sharp, often obscuring frontal suture, often joining anterior to coronal suture; sagittal crest sometimes prominent; supraoccipital shield small, not projecting far posteriorly; external side of occipital well ossified; tympanic bullae usually small, not much inflated; medial part of posterior margin of palate often extending well behind toothrow; mandible thick, deep, ventral edge often convex when viewed from side, toothrow bowed out prominently in center; ascending ramus sometimes curving dorsoposteriorly; teeth relatively small, usually lacking trenchant cusps, usually widely spaced in jaws; upper canines thick anteroposteriorly, short dorsoventrally; P4 usually lacking prominent deuterocone and lingual cingulum; M1 having relatively large paracone and metacone, relatively small medial section, usually lacking buccal cingulum; M2 usually small; p2 and p3 sometimes with posterior cusps; p4 with second cusp, sometimes lacking third cusp, usually without posteromedial cingulum extending behind third cusp; m1 relatively broad, usually with relatively small talonid; m3 occasionally absent. For details on pelage and postcranial skeleton see Haag (1948); Iljin (1941); Miller, Christensen, and Evans (1965); and Scott and Fuller (1965).

Comparison with C. latrans.—Often close in size; skull usually relatively broader in most dimensions; rostrum relatively broader and deeper (a few specialized breeds, such as collies and Russian wolfhounds, may have relatively longer and narrower rostra than coyotes); braincase relatively smaller, less inflated dorsoposteriorly, broader ventrally, set lower relative to other parts of skull (some breeds with smaller skulls than coyotes, may have relatively well inflated braincases, broader at level of parietotemporal sutures than at base); postorbital constriction narrower, more elongate, rising much more steeply into frontal region; zygomata thicker, deeper, usually more broadly flaring; orbits usually relatively smaller; frontals rising much more steeply above rostrum, more depressed medially, more prominently convex, and forming broader, more bulging shield; orbital angle greater; temporal ridges usually sharper; more often obscuring frontal suture, more often joining anterior, rather than posterior to coronal suture; sagittal crest often more prominent (may be absent in some smaller breeds), usually sloping more posteroventrally; supraoccipital shield often not projecting so far posteriorly; external side of occipital more ossified, more often lacking thin-walled projection dorsal to foramen magnum (present in some small individuals); tympanic bullae relatively smaller, usually more rugose and much less inflated; palate relatively broader, central part of posterior margin sometimes extending well behind toothrow; distance between toothrow and bulla relatively longer; mandible relatively thicker and deeper, ventral margin sometimes more convex when viewed from side, toothrow more bowed outward in center; teeth usually relatively smaller with less trenchant cusps, more widely spaced in jaws (some specimens have crowded teeth); upper canines relatively thicker anteroposteriorly and much shorter dorsoventrally; P4 more often lacking prominent deuterocone and lingual cingulum; M1 having relatively larger paracone and metacone, relatively smaller medial section, more often lacking buccal cingulum; M2 relatively much smaller; p2 more often having posterior cusp; p4 having posterior cusps more reduced, more often lacking well developed third cusp and posteromedial cingulum; m1 relatively broader, metaconid less prominent and not projecting so far medially; m2 and talonid of m1 usually relatively smaller with less trenchant cusps; dental and other anomalies more common. Atkins and Dillon (1971) listed differences between *C. familiaris* and *C. latrans* in the morphology of the cerebellum.

Comparison with C. lupus.—Usually much smaller; skull usually much smaller (some breeds, as Irish wolfhounds and great Danes, may have skulls larger than those of most wolves); rostrum usually relatively shorter; braincase usually broader based, set relatively lower; postorbital constriction usually relatively broader and rising more steeply into frontal region; zygomata usually shallower, not so broadly flaring; frontals usually rising more steeply above rostrum, more depressed medially, more prominently convex, and forming relatively higher, broader, and more bulging shield; orbital angle usually greater; temporal ridges more often joining posterior, rather than anterior to coronal suture; sagittal crest usually less prominent (may be higher in some large breeds); supraoccipital shield much smaller, not projecting so far posteriorly; tympanic bullae usually smaller, less inflated; central part of posterior margin of palate more often extending behind toothrow (especially in wolf-sized individuals); distance between bulla and toothrow usually relatively larger; mandible usually relatively thicker, ventral edge more often convex when viewed from side (especially in wolf-sized individuals), toothrow usually more bowed outward in center; teeth, including incisors, of larger individuals relatively much smaller and more widely spaced in jaws; dental and other anomalies more common. Atkins and Dillon (1971) listed differences between *C. familiaris* and *C. lupus* in the morphology of the cerebellum; Hildebrand (1952b), Iljin (1941: 377-379), and Young (1944:179) discussed distinguishing features of external appearance.

Remarks.—An account of *C. familiaris* is herein included primarily for comparative purposes. The origin of the domestic dog, and its fossil and archeological history in North America, are not within the scope of this paper. Most authorities now think that the domestic dog was derived from one of the small Eurasian subspecies of *C. lupus* (Degerbøl, 1961; Lawrence, 1966; Olsen and Olsen, 1977; Reed, 1961; Scott, 1968; Trouessart, 1911). Skaggs (1946:345) suggested that Indian dogs found at a site in Kentucky may have had a coyotelike ancestor, but Allen (1920:440) thought that all American aboriginal dogs had been introduced from the Old World. The oldest known remains of *C. familiaris* in North America, dated at about 10,400-11,500 B.P., were obtained from Jaguar Cave in Lemhi County, Idaho (Lawrence, 1966, 1968).

Attempts to describe the cranial characters that distinguish *C. familiaris* and *C. lupus* have been common in the literature of several nations for many years. One of the earliest and most detailed of these efforts was that of Serres (1835), who listed the following characters of the dog, as compared to those of the wolf: braincase more broadly based; frontals rising more steeply, with broader postorbital processes; sagittal crest less prominent; supraoccipital shield smaller; canine teeth shorter dorsoventrally; carnassials shorter anteroposteriorly; mandible thicker with more convex ventral surface; m3

TABLE 4

Orbital angle (in degrees) of *C. latrans*, *C. lupus*, *C. familiaris*, and *C. dirus*.

	mean	lower extreme	upper extreme	sample size
C. latrans, western U.S.	42.8	36	50	53
C. lupus, western U.S.	42.8	38	49	76
C. familiaris	52.9	40	64	58
C. dirus, Rancho La Brea	53.1	48	60	75

occasionally lacking. A few other features mentioned by Serres do not now seem as reliable as those listed here.

Reynolds (1909:22-23) compiled a list of cranial characters, mostly those cited earlier by Serres, that had been used by various authors to separate the dog and wolf. He reported the most important character to be that the plane of the eye socket is more obliquely inclined to the brow (the orbitofrontal or orbital angle is less) in the wolf. The method of measuring this angle was illustrated by Iljin (1941:387) and Mech (1970:27). I took the measurement on many specimens and found it to be reliable in separating skulls of dogs from those of both wolves and coyotes (see table 4).

Other characters suggested as useful in distinguishing skulls of dogs, as compared to those of wolves, are: smaller average size, relatively smaller teeth, less inflated bullae (Miller, 1912c:313); projection of palate behind M2 (Allen, 1920:436); less acute angle of zygomatic process of maxillary, smaller bullae (Iljin, 1941:390); braincase more ossified and joining rostrum at greater angle (lower set relative to other parts of skull), interorbital region more elongated (Lawrence and Bossert, 1967:225); mandible relatively thicker lateromedially (Lawrence, 1968).

The cranial differences between *C. latrans* and *C. familiaris* received less attention until relatively recently. Coues (1873) and Packard (1885) were both struck by what they considered to be close resemblance in external appearance of the coyote and the American Indian dog. But Allen (1920:434-435, 450) pointed out that no one had yet made a careful comparison of dog and coyote skulls. He reported that dogs had a smaller heel of m1, less trenchant cusps on the molars and premolars, and a more prominent outward bend of the lower toothrow at the junction of the molar and premolar series.

Other characters proposed as useful in distinguishing skulls of domestic dogs, as compared to those of coyotes, are: more inflated frontal sinuses, greater orbital angle, more obtuse angle in maxillo-jugal suture, smaller tympanic bullae, more widely spaced teeth, smaller medial section of M1 (Hall, 1943); broader rostrum and more widely spaced incisors (Burt, 1946:61-62); smaller ratio obtained by dividing the distance between the inner margins of the alveoli of P1 by the distance from the anterior margin of the alveolus of P1 to the posterior margin of the alveolus of M2 (Howard, 1949); canine teeth shorter dorsoventrally and thicker anteroposteriorly (Jackson, 1951:242); relatively broader base of braincase and deeper mandible (Bee and Hall, 1951); deeper and thicker mandible (Lawrence, 1968).

A problem associated with attempts to distinguish between *C. lupus* and *C. familiaris* is the role of domestication in modifying the phenotype of the skull. The most critical question is whether nutritional or other factors involved in captivity cause the developing skull of a wolf to take on characters normally found in the dog. Studies have demonstrated that wolves raised in captivity sometimes show such changes as a shortening of the jaws, overlapping of the teeth, more steeply raised forehead, and general decrease in size. This subject was discussed in detail by Degerbol (1961), Iljin (1941:390-392), Lawrence (1966), Scott (1968:246-247), and various other authors cited therein. Thus far all of the investigations have concerned what happens to wolves in captivity, but it has not yet been determined whether *C. familiaris* could take on wolflike characters under feral conditions.

Interbreeding of domestic dogs with wolves or coyotes has occasionally occurred, and was discussed in the first main part of this paper. Hybridization between *C. lupus* and *C. familiaris* has long been accepted (see reviews by Allen, 1920:433-434; Iljin, 1941: 360-361; and Young, 1944:180-210). Some early naturalists, such as Coues (1873) thought hybridization between dog and coyote to be common in the west, but Allen (1920) refuted this view on the basis of cranial evidence. Later, however, specimens began appearing in the eastern states, that were considered to represent hybrids between the two species (coy-dogs). Mengel (1971) has reviewed this subject in detail.

Canis dirus Leidy

1854. *Canis primaevus* Leidy, Proc. Acad. Nat. Sci., Philadelphia, 7:200. Not *C. primaevus* of Hodgson, 1833.
1858. *Canis dirus* Leidy, Proc. Acad. Nat. Sci., Philadelphia, 1858, p. 21. Type from banks of Ohio River below Evansville, Vanderburgh County, Indiana.
1869. *Canis indianensis* Leidy, Jour. Acad. Nat. Sci., Philadelphia, 7:368. An inadvertant renaming of *C. dirus*.
1876. *Canis mississippiensis* J. A. Allen, Amer. Jour. Sci., ser. 3, 11:49. Type from Lead Region of Upper Mississippi.
1884. *Canis lupus,* Cope and Wortman, Ann. Rept. State Geol. Indiana, 14:9.
1912. *Canis dirus,* Merriam, Mem. Univ. California, 1:218.
1916. *Canis ayersi* Sellards, Ann. Rept. Florida Geol. Surv., 8:152. Type from Vero, Indian River County, Florida.
1918. *Aenocyon dirus,* Merriam, Univ. California Publ. Bull. Dept. Geol., 10:533.
1918. *Aenocyon ayersi,* Merriam, Univ. California Publ. Bull. Dept. Geol., 10:533.
1929. *Canis (Aenocyon) ayersi,* Simpson, Bull. Amer. Mus. Nat. Hist., 56:572.
1946. *Canis (Aenocyon) dirus,* Stock, Lance, and Nigra, Bull. S. California Acad. Sci., 45:109.
1962. *Canis ayersi,* Weigel, Florida Geol. Surv. Spec. Publ., no. 10, p. 37.
1972. *Canis dirus,* Kurten and Anderson, Tebiwa, 15:37.
1974. *Canis dirus,* R. A. Martin, *in* Pleistocene mammals of Florida (edit. Webb), p. 73.

Type.—Left maxillary fragment with P2-M2; no. 11614, Acad. Nat. Sci., Philadelphia; banks of Ohio River below Evansville, Vanderburgh County, Indiana.

Geological distribution.—Rancholabrean to early Recent.

Geographical distribution.—Known from Alberta, Arizona, Arkansas, California, Florida, Idaho, Illinois, Indiana, Kansas, Kentucky, Louisiana, Missouri, Nebraska, Nevada, New Mexico, Oklahoma, Oregon, Pennsylvania, Tennessee, Texas, Utah, Virginia, West Virginia, Wisconsin, Aguascalientes, Jalisco (possibly), Estado de Mexico, Nuevo Leon, Puebla, and northern Peru.

Description.—Size large for the genus; skull averaging largest in genus, with mostly broad proportions; rostrum elongated, relatively broad and deep; braincase relatively small, not much inflated dorsoposteriorly; postorbital constriction elongated, narrow lateromedially, rising steeply into frontal region; zygomata thick, deep, broadly flaring; orbits relatively small; frontals usually well elevated above rostrum, moderately convex, forming broad shield; temporal ridges usually not sharp, seldom obscuring frontal suture, usually joining at or anterior to coronal suture; sagittal crest prominent, sharp dorsally; supraoccipital shield narrow, projecting far posteriorly; external side of occipital well ossified; tympanic bullae usually moderate in size, not well inflated; vertical plates of palatines flaring broadly anteriorly; posterior end of vomer usually extending only slightly behind posterior nasal opening; postpalatine foramina usually opposite posterior ends of P4; optic foramen and anterior lacerated foramen normally close together in common pit; mandible thick and deep, ventral margin not convex when viewed from side, toothrow bowed outward in center; incisors relatively large; upper canines prominent, thick anteroposteriorly; premolars relatively broad; P4 relatively large, usually lacking prominent deuterocone and lingual cingulum; M1 having relatively large paracone and metacone, relatively small medial section without trenchant cusps, reduced hypocone with ridge seldom extending anteriorly around base of protocone, and sometimes with pronounced buccal cingulum; M2 relatively small; p2 usually lacking posterior cusp; p3 often with second and third cusps; p4 usually having second and third cusps, and pronounced posteromedial cingulum extending behind third cusp; m1 relatively large, usually having relatively small, low-set talonid. For additional details, and description of postcranial skeleton, see Merriam (1912); Stock, Lance, and Nigra (1946); Nigra and Lance (1947); Stock and Lance (1948); and Galbreath (1964).

Comparison with C. lupus.—Skull usually larger; posterior part of rostrum usually not so deep; postorbital constriction rising more steeply into frontal region; frontals less convex, less depressed medially, forming relatively broader and flatter shield; orbital angle greater; temporal ridges smoother, less often obscuring frontal suture; sagittal crest usually more prominent; supraoccipital shield narrower, projecting farther posteriorly, more often having posteroventral hook; vertical plates of palatines flaring more broadly anteriorly; posterior end of vomer not extending so far behind posterior nasal opening; postpalatine foramina usually set more posteriorly; optic foramen and anterior lacerated foramen closer together; mandible usually relatively thicker and deeper; upper canines usually relatively smaller; M1 having more reduced hypocone, its anterior ridge much less often extending anteriorly around base of protocone, and more often with buccal cingulum; p2 more often lacking posterior cusp; p4 with posterior margin of second cusp usually sloping anteroventrally rather than posteroventrally, more often with third cusp and posteromedial cingulum extending behind third cusp; m1 usually relatively larger, usually with relatively smaller, narrower, lower set talonid. For additional details, and comparison of postcranial elements of *C. dirus* and *C. lupus*, see Merriam (1912); Stock, Lance, and Nigra (1946); and Stock and Lance (1948).

Other comparison.—See account of *C. armbrusteri*.

Remarks.—The nomenclatural confusion regarding the type specimen, and other early reported material of dire wolves, was resolved by Merriam (1912:218-221). He correctly determined that *Canis dirus* was the proper name for all specimens that had been reported up to the time of his study.

In his original description of *Canis primaevus* (=*Canis dirus*), Leidy (1854:200) noted: "Certain naturalists may regard the fossil as an indication of a variety only of the *Canis lupus*, and of the correctness of such a view I shall not attempt to decide." Subsequently, Cope and Wortman (1884:10) wrote that "it is impossible to admit this fossil to the rank of a distinct and well defined species, but it appears, in our judgement, to be but a variety which has a living representative in the mountains of Oregon, today." Nonetheless, Leidy's original recognition of the distinctness of the dire wolf was borne out when a large number of specimens became available.

Rancho La Brea has yielded by far the greatest amount of dire wolf material. In fact, there are few complete skulls from all other localities combined. Thus, for purposes of statistical comparison, I have restricted my sample of *C. dirus* to 62 specimens from Rancho La Brea that were complete enough for use in multivariate analysis. These were all unknown as to sex, and so were tested against the combined male and female samples of each subspecies of *C. lupus* for which more than five specimens were available (a total of 467 specimens). Figure 52 shows the resulting ranges of variation of each group, and the complete separation of the dire wolf and the gray wolf. Measurements of the series of *C. dirus* are listed in appendix B (part 15), and their means are compared with those of *C. lupus* in figure 53. The skull of the dire wolf is seen to be much larger than that of *C. lupus*, and to be proportionally broader at the canines and frontal shield, but not so deep, relatively, between the toothrow and orbit. The P4 is proportionally much longer, but the upper canine and M2 have relatively smaller diameters.

The lack of multivariate overlap between *C. dirus* and *C. lupus*, and the striking differences in size and proportion, lead me to no other conclusion than that the two must be treated as distinct species. There also is no statistical evidence to suggest that the dire wolf is ancestral to the living gray wolf.

Although the material from Rancho La Brea is by far the most abundant from any one locality, well preserved upper cranial elements from other sites in California, Mexico, Texas, Missouri, and Kentucky has confirmed that the same species, with the same well marked characters, was broadly distributed in the late Pleistocene. Among the most reliable cranial characters that distinguish *C. dirus* from other wild species of *Canis* are: large over-all size, relatively broad frontal shield, large orbital angle (see table 4), narrow supraoccipital shield projecting far posteriorly, vertical plates of palatines flaring broadly anteriorly, postpalatine foramina set relatively far posteriorly, optic foramen and anterior lacerated foramen close together in common pit, large carnassial teeth, and re-

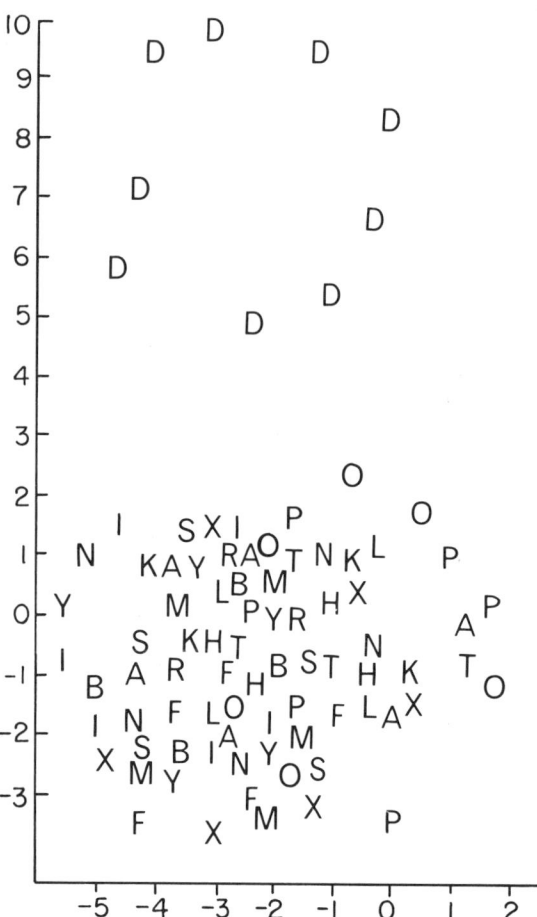

Fig. 52.—Graphical results of multivariate analysis comparing a group of 62 skulls of *C. dirus* from Rancho La Brea (D), with a total of 467 Recent *C. lupus* divided into the following subspecific groups: *C. l. arctos* (A), *C. l. baileyi* (B), *C. l. fuscus* (F), *C. l. hudsonicus* (H), *C. l. irremotus* (I), *C. l. mackenzii* (K), *C. l. ligoni* (L), *C. l. mogollonensis* (M), *C. l. nubilus* (N), *C. l. occidentalis* (O), *C. l. pambasileus* (P), *C. l. lycaon* (R), *C. l. monstrabilis* (S), *C. l. tundrarum* (T), *C. l. bernardi* (X), *C. l. youngi* (Y). As in previous multivariate analyses in this paper, 15 measurements of the skull were used, but in this case sexes were combined in each group. Only marginal positions of individuals of each group are plotted.

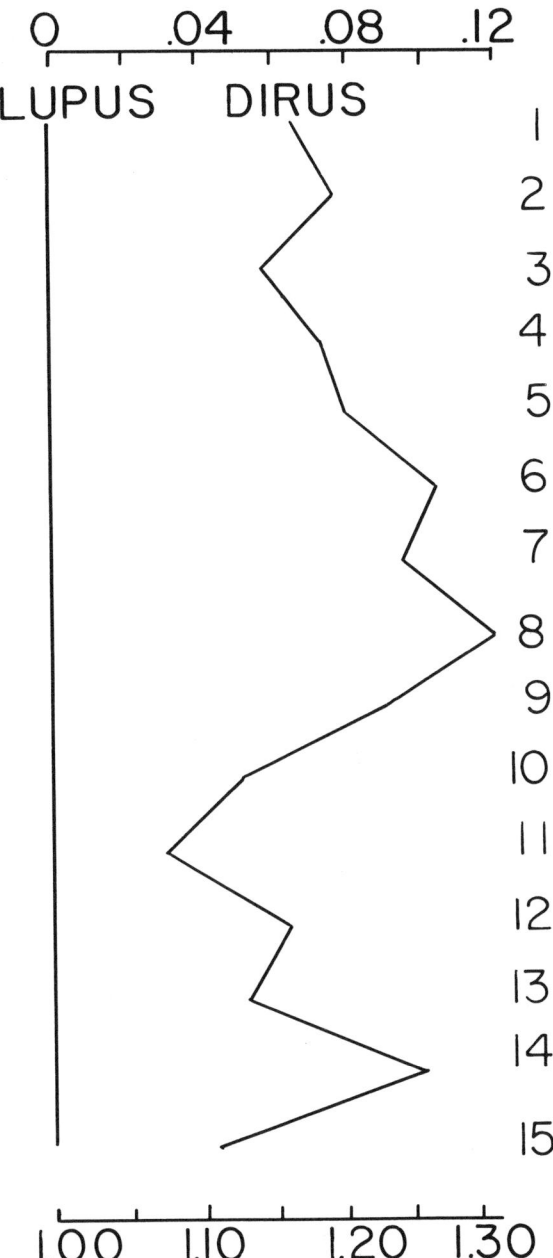

Fig. 53.—Ratio diagram comparing means of *C. dirus* (n=62) and the total sample of Recent *C. lupus* (combined sexes, n=482). Vertically arranged numbers represent the measurements so numbered in appendix B. A log difference scale is provided above, and a ratio scale below the diagram.

duced hypocone on M1. I examined more than 500 skulls from Rancho La Brea in the Los Angeles County Museum, and found few that lacked more than one of these characters. There was no difficulty in assigning most of this material to *C. dirus*, and picking out the eight skulls in the LACM collection that are referable to *C. lupus* (see pp. 99-100). The specimens from Rancho La Brea represent a population that was geographically local, but the deposits were laid down over many thousands of years, and thus considerable intraspecific variation would be expected at the site.

Sellards (1916:152-157) described *Canis ayersi* from Vero, Florida as belonging to the same group as *C. dirus*, but being specifically distinct. The single skull on which he based his description was thought to have a relatively narrower rostrum than the skulls found at Rancho La Brea. But the rostrum of *ayersi* had apparently been damaged prior to its recovery, and since Sellards' study some restorative work has been carried out. A full evaluation is now impossible, but the skull is clearly narrower than most specimens from Rancho La Brea. In addition, some of the mandibles from Florida that I examined are longer than any from the southwest. Dire wolves in Florida may therefore have tended to have relatively long and narrow jaws, but the differences are slight and I do not consider them to be of specific value. In all other characters, as acknowledged by Sellards, the Florida material resembles *C. dirus* from Rancho La Brea.

In describing *C. ayersi*, Sellards (1916: 156) noted that there was no effective barrier between Florida and Evansville, Indiana, the type locality of *C. dirus*, and that the type of *C. dirus*, a maxillary fragment, might eventually be shown to represent the same species as found in Florida. Olson (1940:44) suspected that the Indiana and Florida specimens were probably referable to the same species, *C. dirus* on the basis of nomenclatural priority, and that a new name might be needed for material from California. Recently, a very broad skull of a dire wolf was obtained at Welsh Cave, Woodford County, Kentucky, only 150 miles from Evansville, and was assigned to *C. dirus* by Guilday, Hamilton, and McCrady (1971:274). In rostral proportions this specimen resembles skulls from Rancho La Brea more than it does the type of *C. ayersi*, and it thus supports recognition of the type of *C. dirus* and specimens from Rancho La Brea as representing the same species.

Martin (1974:73) reported that a skull from Reddick, Marion County, Florida was similar in proportion to specimens from Rancho La Brea, and he thus implied that there was morphological overlap between the two populations represented. Martin suggested that the type of *C. ayersi*, as well as all other dire wolf material from Florida, was referable to *C. dirus*, and I have followed the same course in this paper. I have not retained *ayersi* as a subspecies, and presently prefer to recognize *C. dirus* as a widespread monotypic species.

In a popular article, Frick (1930:79) listed two new subspecific names, *Aenocyon dirus alaskensis* from the pre-tundra fauna of Alaska-Yukon, and *A. dirus nebrascensis* from the Sheridan fauna of Nebraska. No type specimens or exact localities were designated, and no descriptions were provided. Nonetheless, several authors referred to supposed dire wolves from Nebraska as *nebrascensis* Frick (Schultz and Stout, 1948:565; Schultz and Tanner, 1957:71; Schultz and Martin, 1970:347). According to Schultz (1934:369), there were specimens of *A. d. nebrascensis* Frick from Hay Springs, Sheridan County, Nebraska in the Frick collection of the American Museum of Natural History. No one, however, discussed particular specimens, and the original faunal list from this site had included only the gray wolf (Matthew, 1918). The specimens of wolves from Hay Springs that I examined are referable to *C. lupus*, and are discussed above in the account of that species. The only specimens of wolves reported from specific localities in Alaska also have been referred to *C. lupus*, and it is unlikely that the dire wolf ever occurred so far north. I therefore consider Frick's names as *nomena nuda* and have not employed them in this paper.

Merriam (1918) created the genus *Aenocyon* to include what he thought were three species: *A. dirus*, *A. ayersi*, and *A. milleri*. Goldman (1944:400) observed that the cranial and dental details of the dire wolf justified recognition of *Aenocyon* as a genus or subgenus. Stock, Lance, and Nigra (1946:109) formally used *Aenocyon* as a subgenus for *C. (A.) dirus*. Various other authors have either followed Merriam, employed *Aenocyon* as a subgenus, or ignored the term (see "Record of occurrences," below).

As pointed out above, *milleri* and *ayersi* are synonyms of *C. lupus furlongi* and *C. dirus*, respectively, and thus there is only one species of dire wolf, *C. dirus*. Although this species has the most pronounced specific characters of any North American member of the genus *Canis*, these characters do not seem to be of a higher order than those distinguishing other species. Furthermore, there is no evidence that the lineage of the dire wolf was long separate from that of other *Canis*. Indeed, *C. dirus* seemingly was the last wild species of *Canis* to appear in North America, and it probably descended from the same basic stock that gave rise to other wolves. Thus I have not employed *Aenocyon* as a genus or subgenus, and have synonymized it under *Canis*.

Record of occurrences.—The dire wolf is represented by more fossil material than any other species of *Canis*, and yet its known geological range is relatively short, being restricted to the Rancholabrean and early Recent. A number of large canids from older sites were assigned to *C. dirus* by previous authors, but in such cases identification was incorrect or material was inadequate to allow careful evaluation. Irvingtonian records from Irvington, California (Savage, 1951:230); Rushville fossil quarry (Schultz and Tanner,

Fig. 54.—Map showing localities (black dots) of *C. dirus*. Because of the scale of the map, it was not possible to plot all localities in crowded areas.

1957:71) and Angus fossil quarry (Schultz and Martin, 1970:347), Nebraska; Port Kennedy, Pennsylvania (Cope, 1899:227); and Rock Creek, Texas (Troxell, 1915:633) are discussed above in the account of *C. armbrusteri*. And records from Hay Springs (Schultz, 1934:369; Schultz and Stout, 1948: 563; Schultz and Tanner, 1957:71) and Mullen (Martin, 1972:174), Nebraska are discussed above in the account of *C. lupus.* In addition, the dire wolf was associated with the Blancan Broadwater quarry site, Nebraska (Barbour and Schultz, 1937:4), and Blanco fauna, Texas (Vanderhoof, 1937). No particular specimens, however, were discussed, and subsequent papers on the Broadwater quarry (Schultz and Stout, 1945:234; 1948:563; Hibbard, 1970:414) and Blanco fauna (Meade, 1945; Johnston and Savage, 1955:36-37; Dalquest, 1975) did not mention the presence of *C. dirus.* Perhaps the initial reports of *C. dirus* had been based mistakenly on remains of *Borophagus.*

The following list is arranged alphabetically by state and province, and geographically (north to south, west to east) within states and provinces, except that Latin American areas are placed last. Specimens examined by me are identified by element, museum number, or both; and selected measurements are found in appendix B (part 15) and appendix C (part 7). Occurrences also are shown on the map in figure 54.

ALBERTA.—Castleguard icefield, Banff National Park; late Pleistocene; as *C. dirus* (Cowan, 1954:44). This record is based on a single lower canine tooth found lying on the surface.

Medicine Hat; Sangamon; as *C. dirus* (Churcher, 1970:63).

ARIZONA.—Ventana Cave, Papago Indian Reservation, Pima County; Wisconsin (Hibbard, 1958); as *C. dirus* (Colbert, 1950:132).

Murray Springs, 1 mi. W Lewis Spring on San Pedro River, Cochise County; late Pleistocene; two mandibular fragments, UAriz 4394, 4395.

Whitewater Draw, near Douglas, Cochise County; early Recent; as *C. dirus* (Hester, 1960:69).

ARKANSAS.—Peccary Cave, eastern Newton County; early Recent; as *C. dirus* (Davis, 1969:164; Quinn, 1972:92).

CALIFORNIA.—Samwel Cave, Shasta Lake, Shasta County; Wisconsin (Hibbard, 1958); as *C. dirus* (Kurten and Anderson, 1972:37); mandibular fragment, UCMP 9566.

Potter Creek Cave, 1 mi. SE Baird, Shasta County; Wisconsin (Hibbard, 1958); as *C. indianensis* (Sinclair, 1904:17), as *C. dirus* (Anderson, 1968:22).

Hawver Cave, 5 mi. E Auburn, El Dorado County; Wisconsin (Hibbard, 1958); as *C.* near *dirus* (Stock, 1918:478); as *C. dirus* (Anderson, 1968:22).

Cool quarry, El Dorado County; late Pleistocene; mandibular fragment, UCMP 38328.

Teichart gravel pit, Sacramento County; late Pleistocene; mandibular fragment, UCMP 85380.

Arroyo Las Positas, Alameda County; Pleistocene; mandible, FM PM664.

Livermore Valley, near San Leandro, Alameda County; Pleistocene; as *C. indianensis* (Leidy, 1873:230), as *C. dirus* (Merriam, 1912:244).

Oil Springs, Tulare County; Pleistocene; as *C. indianensis* (Merriam, 1903:288), as *C. dirus* (Merriam, 1912:244).

McKittrick tar seeps, Kern County; Wisconsin; as *A. dirus* (Merriam and Stock, 1921; Schultz, 1938b:169); two complete skulls, LACM; six maxillary fragments, LACM; 13 mandibles, LACM; three mandibles, UCMP.

Maricopa Brea, near Maricopa, Kern County; Wisconsin (C-14 date: 13,860 B.P.); as *C. dirus* (Shakespear, 1975); three complete skulls with mandibles, six other mostly complete skulls; 10 cranial and maxillary fragments; 20 mostly complete mandibles; many other fragments and isolated teeth, all in LACM.

Carpinteria asphalt, Santa Barbara County; Wisconsin (Hibbard, 1958); as *A.* near *dirus* (Wilson, 1933:69).

Rancho La Brea, Los Angeles, Los Angeles County; Wisconsin; as "*C. indianensis*(?)" (Merriam, 1906), as *C. dirus* (Merriam, 1912:218; Marcus, 1960:2), as *A. dirus* (Merriam, 1918:533; Stock, 1929:286), as *C. (A.) dirus* (Stock, Lance, and Nigra, 1946:109; Nigra and Lance, 1947:26; Stock and Lance, 1948:79); 520 complete or mostly complete skulls without associated mandibles, LACM; one skull, AMNH; two skulls, KU; five skulls, USNM; ten skulls, UCMP; 90 complete mandibles, LACM; three mandibles, AMNH; three mandibles, USNM; five mandibles, UCMP; numerous fragments and isolated teeth, LACM. The dire wolf material from Rancho La Brea is the most abundant of any large mammal from any fossil site in North America. The number of individual wolves represented in the collection of the Los Angeles County Museum was estimated at 2,000 by Stock, and was counted at 1,646 by Marcus (1960). The amount of well preserved material offers an unparalleled opportunity for studies of variation in a local population of canids, but as yet little has been done in this regard. Nigra and Lance (1947) found the average size of metapodials of *C. dirus* to differ between the five major tar pits from which remains were recovered. Population studies of this kind are qualified by the probability that the different pits were active at different times, and that each was active over a lengthy period in which chronological mixing of elements occurred. Radiocarbon dates based directly on specimens of *C. dirus* from Rancho La Brea were given as 9,860±550 and 10,710±320 B.P. by Miller (1968:14). It is likely, however, that remains of wolves were deposited at the site over thousands of years. I found the skulls of *C. dirus* to be remarkably consistent in the critical characters that distinguish the species, and on the whole I agree with the detailed description provided by Merriam in 1912. Studies of postcranial elements from Rancho La Brea have indicated that the dire wolf had a relatively stockier body, lighter limbs, and shorter feet than modern *C. lupus* (Merriam, 1912:236; Stock, Lance, and Nigra, 1946; Stock and Lance, 1948:79).

Harold Beds, 5 mi. SE Palmdale, Los Angeles County; Pleistocene; three metapodials, USNM 13085.

Harbor freeway, Los Angeles, Los Angeles County; Wisconsin; as *C.* cf. *dirus* (Miller, 1971:54).

San Pedro, Los Angeles County; Wisconsin; as *C.* cf. *dirus* (Miller, 1971:45).

La Mirada, Los Angeles County; Wisconsin; as *C.* cf. *dirus* (Miller, 1971:49).

Newport Bay Mesa, Orange County; Wisconsin; as *C.* cf. *dirus* (Miller, 1971:34).

Costeau pit, 2 mi. S El Toro, Orange County; Wisconsin; as *C.* cf. *dirus* (Miller, 1971:17).

FLORIDA.—Aucilla River IA, Jefferson County; Wisconsin; as *C. dirus* (Webb, 1974b:17).

Ichetucknee River, Columbia County; Wisconsin; as *C. dirus* (Webb, 1974b:17); maxillary fragment, UF 8006; three mandibular fragments, UF 8005, 12899, 17717; three canine teeth, UF 1995, 8214, 8215; postcranial fragments, MCZ 18347-18349. The mandibles resemble those of dire wolves from Rancho La Brea, but are larger.

Santa Fe River IIA, Gilchrist County; Rancholabrean (Webb, 1974b:31; apparently incorrectly designated as late Irvingtonian on p. 13); as *C. dirus* (Webb, 1974b:17).

Hornsby Springs, near High Springs, Alachua County; Wisconsin; as *A. ayersi* (Bader, 1957:71); maxillary fragment, UF 3988; mandibular fragment, UF 3987.

Haile VIIIA, Alachua County; Sangamon; as *C. dirus* (Webb, 1974b:17).

Arredondo IB, 4 mi. SW Gainesville, Alachua County; Sangamon; as *A. ayersi* (Bader, 1957:54), as *C. dirus* (Webb, 1974b:17).

Devil's Den, near Williston, Levy County; late Wisconsin or early Recent (7,000-8,000 B.P.); as *C. dirus* (Martin and Webb, 1974:126); incomplete skull, UF 7996.

Wekiva River, Levy County; late Pleistocene; mandibular fragment, UF 14204.

Reddick IA, Marion County; Sangamon (Webb, 1974b:13); as *C. ayersi* (Gut, 1939), as *C. (A.) ayersi* (Gut and Ray, 1964), as *C. dirus* (Martin, 1974:73); crushed skull with mandibles, UF 2923; two crushed skulls without mandibles, UF 3081, one unnumbered; mandibular fragment, UF; two isolated M1, isolated M2, P4, m1, UF; P4, MCZ.

Eichelberger Cave, 2 mi. SW Belleview, Marion County; late Pleistocene; cast of pair of mandibles, MCZ 7349; two mandibular fragments (probably from same individual), UF 1622, 1623; isolated teeth, UF. The mandibles and lower carnassials are the largest that I examined.

Sabertooth Cave, 1 mi. NW Lecanto, Citrus County; Wisconsin; as *C. ayersi* (Simpson, 1928:9), as *C. dirus* (Webb, 1974b:17).

Rock Springs, Orange County; Sangamon; as *C. dirus* (Webb, 1974b:17).

Seminole Field, near St. Petersburg, Pinellas County; Wisconsin (Hibbard, *et al.*, 1965); as *C. (A.) ayersi* (Simpson, 1929a:572), as *C. dirus* (Webb, 1974b:17); mandibular fragment, AMNH 23568; M1, AMNH 23582; M2, AMNH 23569; two m1, AMNH 23565, 23567; various other fragmentary teeth, AMNH. Simpson reported that a large and a small kind of canid were represented both at Seminole Field and Sabertooth Cave. All of the material that I examined appears to be within the range of variation of *C. dirus*. A small P4 listed by Simpson may possibly have belonged to a large red wolf. Webb (1974b:17), and Martin and Webb (1974: 128) reported *C. familiaris* from Seminole Field. A C-14 date of only 2,040±90 B.P. for this site was questioned by Hester (1960).

Melbourne, Brevard County; Wisconsin (Hibbard, *et al.*, 1965); as *C. (A.)* cf. *ayersi* (Gazin, 1950:400), as *A.* cf. *ayersi* (Ray, 1958:433), as *C. dirus* (Webb, 1974b:17); mandible, USNM 12946; two isolated P4, two M1, four m1, USNM.

Sebastian Canal, Brevard County; Wisconsin; as *C. dirus* (Webb, 1974b:17).

Vero (stratum 2), Indian River County; late Wisconsin (Webb, 1974b:13); as *C. ayersi* (Sellards, 1916:152; Weigel, 1962:37), as *C. dirus* (Martin, 1974:73); skull without mandibles, FGS 7166. The status of *C. ayersi*, the type of which was obtained at Vero, is discussed in the above "remarks."

Bradenton, Manatee County; Sangamon (Webb, 1974b:13); maxillary fragment, UF 3276; mandibular fragment, UF 2259.

Phillipi Creek-Fruitville Ditch, 7 mi. E Sarasota, Sarasota County; Wisconsin; as *C. ayersi* (Simpson, 1929b:275).

IDAHO.—Jaguar Cave, Beaverhead Mountains, Lemhi County; late Wisconsin (C-14 dates: 10,370 ±350 and 11,580±250 B.P.); as *C.* cf. *dirus* (Kurten and Anderson, 1972:24).

American Falls, Power County; Rancholabrean (Hibbard, *et al.*, 1965), Illinoian (Kurten, 1974:7); as cf. *A. dirus* (Gazin, 1935:298), as *C. (A.) dirus* (Hopkins, Bonnichsen, and Fortsch, 1969:3). Gazin's original faunal list for this site stated only that distal portions of two humeri and an abraded phalange could not be distinguished from corresponding parts of dire wolves from Rancho La Brea. This material is not reliable in the identification of *C. dirus*, and might represent some other large canid.

Rainbow Beach local fauna, American Falls Reservoir, Power County; Wisconsin (C-14 dates: 21,500±700 and 31,300±2,300 B.P.); as *C. dirus* (McDonald and Anderson, 1975:26).

ILLINOIS.—Galena, Jo Daviess County; Wisconsin (Kurten, 1974:10); as *C.* or *A. mississippiensis* (Hay, 1923:337).

INDIANA.—Ohio River, below Evansville, Vanderburgh County; late Pleistocene; as *C. primaevus*

(Leidy, 1854:200; 1856:167), as *C. dirus* (Leidy, 1858:21; Merriam, 1912:240), as *C. indianensis* (Leidy, 1869:368), as *C. lupus* (Cope and Wortman, 1884:9), as *A. dirus* (Hay, 1923:204); maxillary fragment, ANSP 11614. The nomenclatural history of the type specimen of *C. dirus* was reviewed by Merriam (1912:218-221). As he noted (1912: 240-241), the type resembles specimens from Rancho La Brea in the reduction of the hypocone of M1, as well as in other features that can be evaluated. There is a pronounced buccal cingulum on the M1, as found in some specimens of *C. dirus,* and the teeth are within the size range of those from California.

KANSAS.—Twelve Mile Creek, Logan County; Pleistocene; as *C. occidentalis* (Hay, 1924:143, 165); P4, M2, p4, KU 392.

Pendennis, Lane County; Pleistocene; as *C. occidentalis* (Hay, 1924:71); m1, KU 393.

Cragin Quarry local fauna, north of Cimarron River, Meade County; Sangamon; as "*C. occidentalis?*" (Hay, 1917b:48), as *A. dirus* (Hibbard, 1939: 464; 1949:84; Hibbard and Taylor, 1960:178), as *C. dirus* (Schultz, 1969:53); mandible, KU 4613.

KENTUCKY.—Welsh Cave, 3.5 mi. SW Troy, Woodford County; late Wisconsin (*ca.* 13,000 B.P.); as *C. dirus* (Guilday, Hamilton, and McCrady, 1971: 274); cast of skull without mandibles, CM 12625; cast of mandible from different individual, CM 12625a. The specimens are among the most complete that have been collected at sites outside of the southwest, and in all characters they resemble specimens from Rancho La Brea.

LOUISIANA.—Avery Island, Iberia Parish; late Pleistocene; as *C. dirus* (Gagliano, 1967:40).

MISSOURI.—Brynjulfson Caves, 6 mi. SSE Columbia, Boone County; late Wisconsin (about 10,000 B.P.); as *C. dirus* (Parmalee and Oesch, 1972:31); isolated teeth, ISM. On the basis of radii, Parmalee and Oesch reported that the wolves from this site were slightly larger than the huge individual found at Powder Mill Creek Cave (see below).

Cherokee Cave, St. Louis, St. Louis County; late Pleistocene (Webster, 1964); as *Canis* (Simpson, 1949:16). Simpson reported that eight metapodials were larger than those of *C. dirus,* but he did not refer them to a species. The measurements he listed are much greater than the means given by Nigra and Lance (1947) for the same elements of *C. dirus* from Rancho La Brea; but the size of more recently collected metapodials of Missouri *C. dirus,* found in association with cranial material, is close to that of the Cherokee Cave specimens (Galbreath, 1964; Hawksley, 1963). Galbreath's specimen (see account of Powder Mill Creek Cave, below) reportedly represented a female, and a male of the same population probably would have had metapodials as large as those found in St. Louis.

Herculaneum (near), Jefferson County; Wisconsin (Hibbard, *et al.,* 1965); as *C. dirus* (Olson, 1940:42); P4, M2, FM WC1736.

Carroll Cave, Camden County; Wisconsin; as *A. dirus* (Hawksley, 1963), as *C. dirus* (Hawksley, 1965:79).

Perkins Cave, Camden County; Wisconsin; as *C. cf. dirus* (Hawksley, 1965:82).

Bat Cave, 8 km. NW Waynesville, Pulaski County; late Wisconsin (10,000-16,000 B.P.); as *A. dirus* (Hawksley, 1963), as *C. dirus* (Hawksley, 1965:81; Hawksley, Reynolds, and Foley, 1973:72-77). According to these last authors, data from Bat Cave tended to bear out Galbreath's (1964) suggestion that dire wolves from Missouri were larger than those from Rancho La Brea (see account of Powder Mill Creek Cave, below).

Cox Cave, Pulaski County; late Pleistocene; "possibly *Canis dirus*" (Mehl, 1962:44).

Bushwacker Cave, Pulaski County; Wisconsin; as *C. dirus* (Hawksley, Reynolds, and Foley, 1973: 73).

Powder Mill Creek Cave, Shannon County; late Wisconsin (C-14 date: 13,170±600 B.P.); as *C. (A.) dirus* (Galbreath, 1964). Galbreath reported the discovery of most of the skeleton, but not including the upper parts of the skull, of a large female dire wolf. Most of the postcranial measurements were found to exceed those of even the largest reported specimens of *C. dirus* from Rancho La Brea (as listed by Merriam, 1912; Nigra and Lance, 1947; and Stock and Lance, 1948). Galbreath thus considered the limbs and feet of the Missouri individual to be relatively larger than those of specimens from the tar pits. The size of this individual, and of others reported by Hawksley (1963), led Galbreath to suggest the possibility that Missouri dire wolves averaged larger than those of California.

Zoo Cave, 1 mi. ENE Hilda, Taney County; late Wisconsin (9,000-13,000 B.P.); as *C. dirus* (Hood and Hawksley, 1975:25, 28). According to these authors the material from this site represented an adult that was "quite small by Missouri standards, closely approaching the size of Rancho La Brea specimens."

NEBRASKA.—Heckendorf gravel pit, Stanton County; late Pleistocene; cranial fragment, UN 2911.

NEVADA.—Gypsum Cave, 16 mi. E Las Vegas, Clark County; late Wisconsin; as "*Canis* or *Aenocyon* sp." (Harrington, 1933:192), as *C. dirus* (Hester, 1960:69).

NEW MEXICO.—Conkling Cavern, near Las Cruces, Dona Ana County; late Pleistocene; mandible, LACM.

Hermit's Cave, east slope of Guadalupe Mountains, Eddy County; late Wisconsin (C-14 dates: 11,850±350 and 12,900±350 B.P.); as *C. dirus* (Schultz, Martin, and Tanner, 1970); maxillary fragment, UN 19212; cranial fragment, UN 19215; mandibular fragment, UN 19213.

OKLAHOMA.—Marlow, Stephens County; late Pleistocene; skull with mandibles, USNM 10278.

OREGON.—Willamette Valley, near Woodburn, Marion County; late Pleistocene; as *A. dirus* (Packard, 1950:89).

Fossil Lake, Lake County; early or middle Wisconsin (Allison, 1966:32); as *C.* cf. *dirus* (Elftman, 1931:5).

PENNSYLVANIA.—Frankstown Cave, Blair County; Wisconsin (Hibbard, 1958); as *C. dirus* (Peterson, 1926:282); maxillary fragment, CM 11023; three mandibular fragments, CM 11022, 11024, 11026. Although the Frankstown material is referable to *C. dirus*, there is some approach in dental characters to *C. armbrusteri* from Cumberland Cave, only about 50 miles away. More accurate age estimates of these two sites would be desirable, so that we might evaluate the idea of the Frankstown specimens representing a transition between *C. armbrusteri* and *C. dirus*.

TENNESSEE.—Jewell Cave, near Ruskin, Dickson County; late Pleistocene; as "wolf" (Barr, 1961:178), as *C. dirus* (Corgan, 1976).

Robinson Cave, 8 mi. SW Livingston, Overton County; Wisconsin; as *C. dirus* (Guilday, Hamilton, and McCrady, 1969:60).

Whitesburg, Hamblen County; late Pleistocene; as "*A. ayersi?*" (Hay, 1921:95); isolated incisors and premolars, USNM 8997.

TEXAS.—Tule Canyon, Briscoe County; Pleistocene; as *C. indianensis* (Cope, 1895:453), as *C. dirus* (Merriam, 1912:242).

Slaton quarry, 5 mi. N Slaton, Lubbock County; Sangamon (Hibbard, 1970); as *Aenocyon* sp. (Dalquest, 1967:10).

Pemberton Hill, Denton County; Sangamon; as *Aenocyon* sp. (Slaughter, *et al.*, 1962:17).

Moore Pit local fauna, Dallas, Dallas County; Sangamon; as *A.* cf. *dirus* (Slaughter, 1966b:79).

Williams Cave, southern end of Guadalupe Mountains, Culberson County; late Pleistocene; as *C. dirus* (Ayer, 1936:608).

Scharbauer site, south of Midland, Midland County; Wisconsin; as *C. dirus* or *ayersi* (Wendorf, Krieger, and Albritton, 1955:113).

Clamp Cave, San Saba County; early Recent; as *A. dirus* (Lundelius, 1967:293).

Laubach Cave, Georgetown, Williamson County; Wisconsin (Kurten, 1974:9); as *C.* cf. *dirus* (Slaughter, 1966a:481).

Levi shelter, Travis County; late Wisconsin; as *A. dirus* (Lundelius, 1967:293).

Friesenhahn Cave, near Bulverde, Bexar County; Wisconsin; as "*A. dirus?*" (Hay, 1921:141), as *A. dirus* (Lundelius, 1960:38).

Kincaid shelter, Uvalde County; late Pleistocene; as *A. dirus* (Lundelius, 1967:293).

Blanco Creek, Bee County; late Pleistocene; as *A. ayersi* (Sellards, 1940:1636).

Ingleside gravel pit, San Patricio County; Wisconsin; as *A. dirus* (Lundelius, 1962), as *C. dirus* (Lundelius, 1972:12); skull without mandibles, mandibular fragment, M1, various other fragments, TM. The skull is crushed, but appears to be the largest of *C. dirus* that I examined. Lundelius (1972:12, 20) observed that two skulls from the Ingleside fauna resembled the type of *C. ayersi* more than they did skulls from Rancho La Brea, but that there was no basis for considering *C. ayersi* a separate species from *C. dirus*.

UTAH.—Silver Creek local fauna, 5 mi. N Park City, Summit County; late Sangamon to early Wisconsin; as *C.* cf. *dirus* (Miller, 1976:401).

VIRGINIA.—Clark's Cave, 12 km. SW Williamsville, Bath County; late Wisconsin (less than 10,000 B.P.); as *C.* cf. *dirus* (Guilday, 1977:69).

WEST VIRGINIA.—Rennick, Greenbrier County; late Pleistocene; mandible, CM 24327.

WISCONSIN.—Blue Mounds, Dane County; late Pleistocene; as *C. mississippiensis* (Allen, 1876:49; Hay, 1914:484; 1923:342), as *C. dirus* (Merriam, 1912:221); four limb bones, MCZ 10988-10991. Allen originally referred to this record as being from only the "Lead Region of Upper Mississippi," but Hay (1923:342) restricted the locality to Blue Mounds. Allen thought that the great size of the bones warranted their referral to a distinct species, but he compared them only to a single small individual of *C. lupus*. Hay's continued recognition of *C. mississippiensis* also was based on scanty comparative material. The measurements of length listed by Allen actually fall within the size range of both *C. lupus* and *C. dirus* as given by Stock and Lance (1948:82), but are closer to the means of the latter species. It seems best for now to follow Merriam (1912) in synonymizing *C. mississippiensis* under *C. dirus*.

AGUASCALIENTES.—Cedazo local fauna, near City of Aguascalientes; early Rancholabrean (probably Illinoian); as *C. dirus* (Mooser and Dalquest, 1975:788); four mandibular fragments, Midwestern State University Department of Biology 9781-9784. These specimens, clearly referable to *C. dirus*, provide the only well supported record of a pre-Sangamon dire wolf. Actually, however, there is some question about the age of the fauna, as Mooser and Dalquest (1975:783) had stated: "Early Rancholabrean age (Savage, 1951) is indicated. We think the Cedazo local fauna could be as old as Yarmouthian or as young as Sangamon, but favor Illinoian Age."

JALISCO.—Lago de Chapala; late Pleistocene; as "*Canis* sp., large wolf" (Downs, 1958).

ESTADO DE MEXICO.—Tequixquiac (near); late Pleistocene; as *C. dirus* (Merriam, 1912:243), as *A. dirus* (Furlong, 1925:152; Maldonado-Koerdell, 1955); cast of cranial fragment, UCMP 27615.

NUEVO LEON.—San Josecito Cave, near Aramberri; Wisconsin; as *Aenocyon* (Russell, 1960:541); two skulls, LACM 3106, 9795; 27 cranial and maxillary fragments, LACM; 30 mandibular fragments, LACM. This large amount of material represents a population not differing in characters from that of Rancho La Brea.

PUEBLA.—Valsequillo, near Puebla; late Pleistocene; as *C. (A.) dirus* (Thenius, 1970:59).

PERU.—La Brea, 30 mi. SE Talara (northern part of Peru, not mapped in Fig. 54); late Pleistocene; as *C. (A.) dirus* (Churcher, 1959).

Evolutionary position.—The species *C. dirus* is known only from the late Pleistocene and early Recent, and is the most common fossil wolf of that period. The dire wolf was not an ancestral species, but rather a highly specialized animal, well adapted for life in the megafaunal community of its time. In its large size, broad proportions, large teeth, and other critical characters, it stood on the opposite end of the evolutionary line from the small species of *Canis* of the Pliocene and early Pleistocene. Its initial appearance in the southern part of the continent may be correlated with a northern withdrawal of *C. lupus*. Hay (1927:192) speculated that *C. dirus* was restricted to a more southerly range by the presence of the gray wolf in the north.

Both the geographic and phylogenetic origin of the dire wolf are unknown. Kurten (1968:109) suggested that *C. falconeri*, a large wolf of the early Pleistocene of Europe, might be related to *C. dirus* of the New World. But there is no chronological or geographic evidence to support recognition of a connection between the two, and the measurements listed by Del Campana (1913:220-229) indicate that the skull of *C. falconeri* did not closely approach that of *C. dirus* in size.

Martin (1974:76) suggested that a population represented by *C. armbrusteri* of Cumberland Cave, Maryland may have given rise to *C. dirus*. This view is reasonable in that the disappearance of *C. armbrusteri* in the Illinoian coincided with the initial appearance of *C. dirus* (see account of *C. armbrusteri*). Moreover, some specimens of *C. armbrusteri* approach those of *C. dirus* in size, and the two species share other characters such as in the morphology of the lower premolars. As yet, however, there is no conclusive evidence to indicate immediate relationship between *C. armbrusteri* and the dire wolf.

Apparently *C. dirus* developed exclusively in the Western Hemisphere, and its ancestry probably lies in the basic stock of small wolves represented by *C. edwardii* and *C. rufus*. At some point in the Pleistocene, an element of this stock, comprised of individuals resembling either *C. edwardii*, *C. rufus*, or *C. armbrusteri*, must have become isolated and begun separate evolution. But how was *C. dirus* able to appear suddenly all across North America in the late Pleistocene, with the most distinctive set of characters in the genus *Canis* already fully developed?

One hypothesis that can not now be disregarded is that the dire wolf arose and developed in South America. There are several pieces of evidence to support this idea. First, *C. dirus* has been reported from South America, specifically from the La Brea tar pits near Talara, Peru (Churcher, 1959). The species also is known from several sites in Mexico, and at one time probably was distributed throughout that country and Central America. Intriguingly, the earliest occurrence of the species, that is supported by good evidence, also is among the most southerly (see account of Cedazo local fauna, Aguascalientes, above). The known range of the dire wolf (Fig. 54) suggests a southern, warmth adapted species, in contrast to the boreal *C. lupus*. The primitive stock of small wolves, represented by the living *C. rufus*, also seems to have been warmth adapted to some degree. Factors associated with one of the glaciations may have driven an element of this stock into South America where it eventually evolved into *C. dirus*. Possibly the Sangamon interglacial afforded the opportunity for reinvasion of much of North America. Further evidence is offered by available information on large South American fossil *Canis* (L. Kraglievich, 1928; J. L. Kraglievich, 1952: 63). One specimen in particular, *C. nehringi* from the province of Buenos Aires in Argentina, appears to have points of resemblance to *C. dirus*. L. Kraglievich's photographs and measurements show that the skull of *C. nehringi* shares at least the following characters with *C. dirus*: large size and massive proportions, broad frontal shield, prominent sagittal crest, narrow supraoccipital shield projecting far posteriorly, vertical plates of palatines flaring broadly anteriorly, postpalatine foramina opposite posterior ends of P4, and relatively large carnassial teeth. Of course proof of relationship between *C. dirus* and *C. nehringi* would not in itself establish South American origin for the dire wolf.

Although its origin remains a mystery, the dire wolf was clearly a common mammal of the North American late Pleistocene. The nature of its fossil remains suggests that it was found primarily in open lowlands, and was a predator of its contemporary large herbivores. The extinction of most of this megafauna at the close of the Pleistocene, for any or all of the reasons discussed by Martin and Wright (1967), probably also signaled the end of the dire wolf. An additional factor in the extinction of *C. dirus* may have been a renewed influx of gray wolves following the withdrawal of the Wisconsin ice sheet. A general consensus among authors who have speculated on the behavior of *C. dirus* is that it was a powerful creature, but was slower and possibly not so alert as *C. lupus* (Matthew, 1916; Merriam, 1912:218; Scott, 1937:578; Stock, 1956:32; Stock and Lance, 1948). The dire wolf may not have been so well adapted in the pursuit of the predominantly smaller, swifter herbivores that survived through the Recent, and it may have lost in competition with the gray wolf. The sympatric occurrence of the two species is demonstrated by good cranial material from Rancho La Brea, the Maricoa Brea, San Josecito Cave, and Hermit's Cave. Fossils of both species also have been reported from Fossil Lake, Jaguar Cave, Samwel Cave, Potter Creek Cave, Ventana Cave, Blue Mounds, Brynjulfson Caves, and Medicine Hat.

SUMMARY

Systematic problems in the genus *Canis* center on its paleontological history and on Recent populations in eastern North America. In order to investigate these matters, approximately 5,000 specimens were examined. Many of these were complete skulls, 15 measurements of which were utilized in the BMD07M computer program of multivariate analysis. For statistical purposes, material was separated by sex, except for fossils and specimens of domestic dogs.

The gray wolf (*Canis lupus*) and the coyote (*Canis latrans*) are readily distinguished from one another, and from the domestic dog (*Canis familiaris*). By multivariate analysis, only five skulls of wild *Canis* from northern and western North America appeared to represent hybrids. The remaining 379 specimens of *C. lupus* and 277 of *C. latrans* from these regions, along with a series of 50 *C. familiaris*, were used as standard groups with which to compare individuals taken in the east. The subspecies *C. lupus lycaon*, which has been nearly exterminated in the eastern United States, still survives in the upper Great Lakes region, as well as in southeastern Canada. Nearly all specimens that had been previously identified as *lycaon* showed close statistical affinity to the standard sample of *C. lupus*, and thus were combined with that sample.

Available information indicates that by 1900 the coyote had begun to extend its range to the east and north of the prairies. The subspecies *C. latrans thamnos*, of the north-central United States and southeastern Canada, is statistically close to western *C. latrans*. A few specimens, however, suggest that limited wolf-coyote hybridization has occurred recently in southern Ontario and Quebec, and has allowed introgression of genes from *C. lupus* into *C. latrans*. As a result, the multivariate position of the coyote population now expanding through the northeastern United States is shifted in the direction of the wolf. Hybridization of *C. latrans* and *C. familiaris* also has taken place, but has not had substantial effect on wild species of *Canis*.

In historical time, the red wolf (*C. rufus*) inhabited the region from central Texas to the Atlantic, and from the Gulf Coast to the Ohio Valley and Pennsylvania. The 14 earliest available specimens, from the part of this region that was well separated from the original range of the coyote, show no statistical overlap with the standard samples of 482 *C. lupus* (including 103 *lycaon*), 277 *C. latrans*, and 50 *C. familiaris*. An additional 115 skulls collected from 1919 to 1929 in Arkansas, Louisiana, southern Missouri, and eastern Oklahoma, and previously identified as *C. rufus gregoryi*, have almost the same multivariate distribution. These skulls, plus most of the older specimens, were combined to make a standard red wolf sample of 125 individuals. This sample and the standard coyote sample were used to compare all other southeastern material.

Series of specimens taken prior to 1930 indicate that hybridization between *C. rufus* and *C. latrans* generally was uncommon where their ranges approached or overlapped, except in the Edwards Plateau area of central Texas. Material from that area forms a statistical bridge between the ranges of variation of the two standard samples. Subsequently, as the red wolf became rare, hybridization increased along the Texas coast, and in north-central Texas, eastern Oklahoma, southern Missouri, and Arkansas. This interbreeding apparently allowed introgression of red wolf genes into the expanding coyote population, which by the 1960's had become established in most inland areas of the south-central states. This population is essentially coyotelike, but is shifted statistically in the direction of the red wolf, and contains a few individuals that are phenotypically close to *C. rufus*.

Material collected in the 1960's and 1970's shows that the genetic influence of the red wolf remained strong within 100 miles of the Texas coast. Samples from most localities there fall mainly between the statistical distributions of *C. rufus* and *C. latrans*. Until about 1970, an unmodified population of the red wolf survived in extreme southeastern Texas and probably in adjacent parts of Louisiana.

The genus *Canis* apparently arose by the middle Pliocene (Hemphillian), and its subsequent hypothetical evolution is shown in figure 55. The relationships of *C. cedazoensis* of Mexico are not well understood, but otherwise the North American species can be

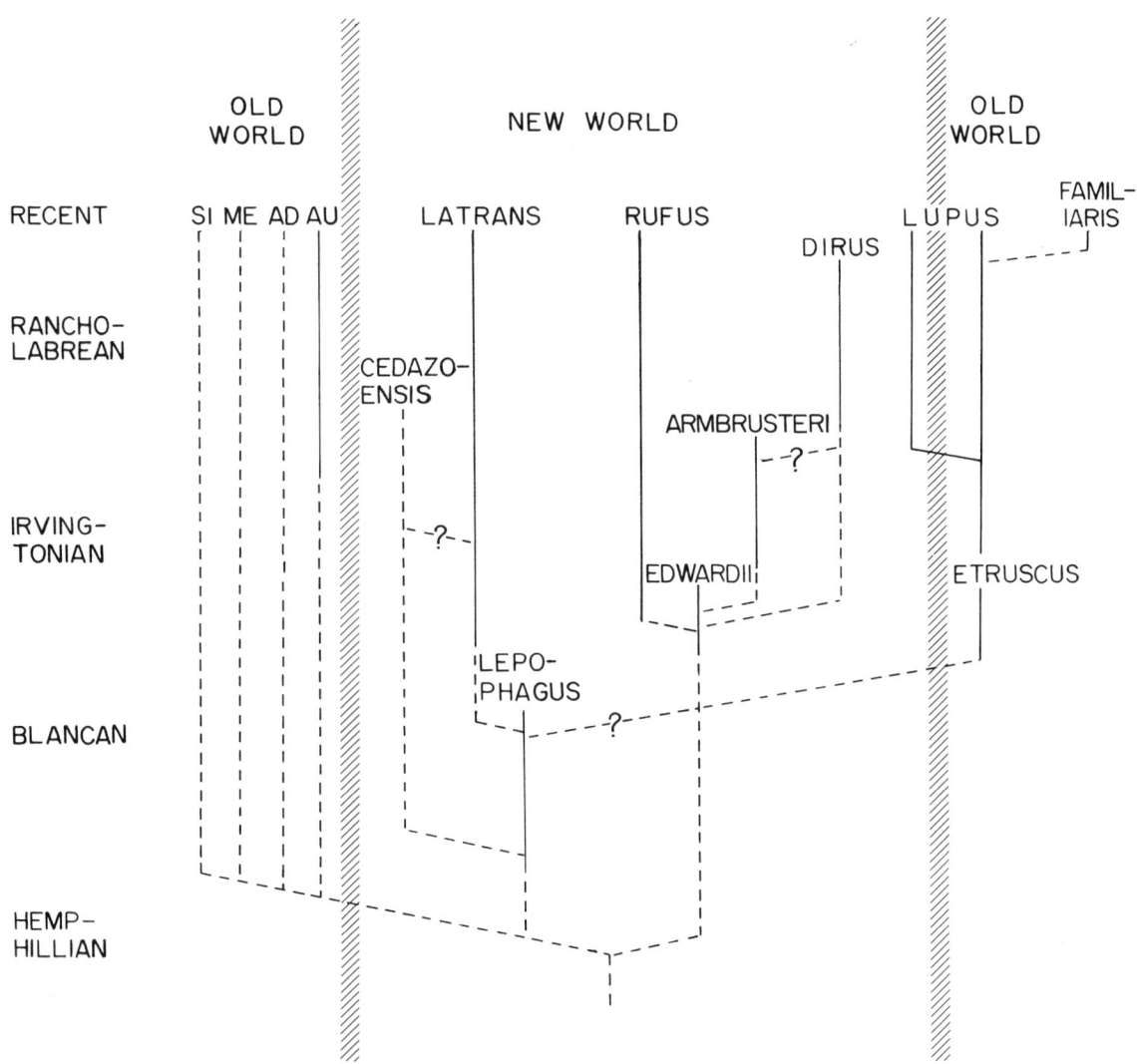

Fig. 55.—Hypothetical phylogenetic diagram of the evolution of *Canis*. Dashed lines indicate possible lineages in which fossil evidence is lacking. Names of species are placed at the latest levels at which those species are known. Only names of species recognized in this paper are shown. Horizontal and vertical distances are not necessarily to scale, and do not indicate degree of affinity. The following species are abbreviated: *C. simensis* (SI), *C. mesomelas* (ME), *C. adustus* (AD), *C. aureus* (AU). Question marks (?) indicate alternative lineages.

separated into coyote and wolf groups. The Blancan *C. lepophagus* was a variable entity, with some individuals resembling small coyotes, and others having certain wolflike characters. It is known from 15 localities, from Florida to Idaho. The species *C. latrans* probably descended from certain populations of *C. lepophagus* by the end of the Blancan, and subsequently there appears to have been little change in the coyote line. Some Pleistocene coyotes, especially those of the late Rancholabrean of California, became larger than modern *C. latrans*, but others were about the same size or smaller. Fossil *C. latrans* has been reported from 109 localities across North America, from Florida to Alaska, and from Oaxaca to Pennsylvania.

It is not clear whether the wolf group descended from some population of *C. lepophagus*, or was already distinct from the coyote line throughout the Blancan. Several species of wolves appeared by the early Irvingtonian, the first of which was probably *C. edwardii*. This was a small species, known from a few localities in the southwestern quarter of North America. An immediate relative, *C. rufus* of the southeast, continued to represent the primitive stock of wolves until Recent times. Still another Irvingtonian wolf, *C. armbrusteri*, was much larger, but resembled the red wolf in certain dental characters and skull proportions. It is known by good material only from Maryland and Florida, though fragmentary remains from elsewhere suggest that it may once have occurred all across the continent.

During the early Pleistocene, an element of the primitive stock of small wolves apparently entered Eurasia where it gave rise to *C. lupus*. In the course of the Illinoian glaciation this species probably moved into North America, where its fossils have been reported from 58 localities. Some Rancholabrean gray wolves were remarkably small, but others, particularly those of Rancho La Brea, had massive skulls resembling those of some modern Arctic wolves.

The extinct *C. dirus* did not appear in North America until the Rancholabrean, and may have originated in South America or descended from *C. armbrusteri*. This large, highly specialized species was not ancestral to modern wolves, and its skull is easily distinguished from that of *C. lupus*. By multivariate analysis, there was no overlap between 62 specimens of *C. dirus* from Rancho La Brea and 467 specimens of Recent *C. lupus*. The dire wolf has been reported from 96 localities and apparently was common throughout that part of North America to the south of Canada.

The generic name *Aenocyon*, sometimes applied to the dire wolf, is here synonymized under *Canis*. The following names no longer are considered to represent separate species: *C. caneloensis*, *C. irvingtonensis*, and *C. riviveronis* (all are fossil subspecies of *C. latrans*); *C. andersoni* (a synonym of *C. latrans orcutti*); *C. petrolei* (a synonym of *C. familiaris*); *C. priscolatrans* (a fossil subspecies of *C. rufus*); *C. milleri* (a synonym of *C. lupus furlongi*); and *C. ayersi* (a synonym of *C. dirus*).

LITERATURE CITED

ALDOUS, C. M.
1939. Coyotes in Maine. Jour. Mamm., 20:104-106.

AKERSTEN, W. A.
1970. Interpretation of sediments and vertebrate fossils in fill of Red Light Bolson, southeastern Hudspeth County, Texas. In Geology of the southern Quitman Mountains area, Trans-Pecos, Texas. Soc. Econ. Paleontol. and Mineral., publ. 70-12, pp. 82-87.
1972. Red Light local fauna (Blancan) of the Love Formation, southeastern Hudspeth County, Texas. Bull. Texas Mem. Mus., no. 20, 53 pp.

ALLEN, G. M.
1920. Dogs of the American aborigines. Bull. Mus. Comp. Zool., 63:431-517.
1942. Extinct and vanishing mammals of the Western Hemisphere. Amer. Comm. Int. Wildl. Prot., xv+620 pp.

ALLEN, J. A.
1876. Description of some remains of an extinct species of wolf, and an extinct species of deer from the lead region of the upper Mississippi. Amer. Jour. Sci., ser. 3, 11:47-51.
1896. On mammals collected in Bexar County and vicinity, Texas, by Mr. H. P. Attwater, with field notes by the collector. Bull. Amer. Mus. Nat. Hist., 8:47-80.

ALLISON, I. S.
1966. Fossil Lake Oregon. Oregon State Univ. Studies in Geol., no. 9, 48 pp.

ALVAREZ, T.
1963. Restos de mamiferos encontrados en una cueva de Valle Nacional, Oaxaca, Mexico. Rev. Biol. Trop., 11:57-61.
1965. Catalogo paleomastozoologico Mexicano. Inst. Nac. Antro. Hist., Mexico, Dept. Prehist., publ. 17, 70 pp.

ANDERSON, E.
1968. Fauna of the Little Box Elder Cave, Converse County, Wyoming. Univ. Colorado Studies, Earth Sci. Ser., no. 6, 59 pp.
1974. A survey of the late Pleistocene and Holocene mammal fauna of Wyoming. Geol. Surv. Wyoming Rept. Invest., 10:76-87.

ANDERSON, R. M.
1943. Summary of the large wolves of Canada, with description of three new Arctic races. Jour. Mamm., 24:386-393.
1946. Catalogue of Canadian Recent mammals. Bull. Natl. Mus. Canada, 102:i-v+1-238.

ARKANSAS GAME AND FISH COMMISSION
1951. A survey of Arkansas game. Little Rock, iv+155 pp.

ARNOLD, D. A.
1952. About wolves. Michigan Conserv., 21(1):23-25.

ATKINS, D. L., AND L. S. DILLON
1971. Evolution of the cerebellum in the genus *Canis*. Jour. Mamm., 52:96-107.

AUDUBON, J. J., AND J. BACHMAN
1851. The quadrupeds of North America. New York, vol. 2, 334 pp.

AYER, M. Y.
1936. The archaeological and faunal material from Williams Cave, Guadalupe Mountains, Texas. Proc. Acad. Nat. Sci., Philadelphia, 88:599-618.

BADER, R. S.
1957. Two Pleistocene mammalian faunas from Alachua County, Florida. Bull. Florida State Mus., Biol. Ser., 2:53-75.

BAILEY, B.
1929. Mammals of Sherburne County, Minnesota. Jour. Mamm., 10:153-164.

BAILEY, H. H.
1930. Correcting inaccurate ranges of certain Florida mammals and others of Virginia and the Carolinas. Bull. Bailey Mus. and Lib. Nat. Hist., no. 5.

BAILEY, V.
1905. Biological survey of Texas. N. Amer. Fauna, no. 25, 222 pp.
1907. Wolves in relation to stock, game, and the national forest reserves. U.S. Dept. Agr. Forest Serv. Bull., no. 72, 31 pp.

BAIRD, S. F.
1857. General report upon the zoology of the several Pacific railroad routes. Part I: Mammals. Washington, D.C., xlviii+757 pp.

BANGS, O.
1898. The land mammals of peninsular Florida and the coast region of Georgia. Proc. Boston Soc. Nat. Hist., 28:157-235.

BARBOUR, E. H., AND C. B. SCHULTZ
1937. An early Pleistocene fauna from Nebraska. Amer. Mus. Novit., no. 942, 10 pp.

BARBOUR, T.
1944. That vanishing Eden. A naturalist's Florida. Little, Brown and Co., Boston, 250 pp.

BARKALOW, F. S., JR.
1976. (Erroneously labeled as 1972.) Vertebrate remains from archeological sites in the Tennessee Valley of Alabama. Southern Indian Studies, 24:3-53.

BARR, T. C.
1961. Caves of Tennessee. Bull. Dept. Conserv. and Commerce, Tennessee, Div. Geol., no. 64, vii+567 pp.

BARTRAM, W.
1791. Travels. Philadelphia, xxxiv+522 pp.

BEE, J. W., AND E. R. HALL
1951. An instance of coyote-dog hybridization. Trans. Kansas Acad. Sci., 54:73-77.

Beezley, C.
1967. Marsh fugitive. Texas Parks and Wildl., 25(1):18-20.

Bennitt, R., and W. O. Nagel
1937. A survey of the resident game and furbearers of Missouri. Univ. Missouri Studies, 12(2):1-215.

Bjork, P. R.
1970. The Carnivora of the Hagerman local fauna (late Pliocene) of southwestern Idaho. Trans. Amer. Phil. Soc., new ser., 60:3-54.

Black, J. D.
1936. Mammals of northwestern Arkansas. Jour. Mamm., 17:29-35.

Black, J. H., and T. L. Best
1972. Remains of a gray wolf (Canis lupus) from northwestern Oklahoma. Proc. Oklahoma Acad. Sci., 52:120.

Bray, W. L.
1904. The timber of the Edwards Plateau of Texas. U.S. Dept. Agr., Bur. Forestry Bull., no. 49, 30 pp.

Bromley, A. W.
1956. Adirondack coyotes. New York State Conserv., 10(4):8-9.

Brown, B.
1908. The Conard fissure. Mem. Amer. Mus. Nat. Hist., 9:157-208.

Bullen, R. P., and C. A. Benson
1967. Cut wolf jaws from Tick Island, Florida. Florida Anthropol., 20:175-177.

Bump, G.
1941. The introduction and transportation of game birds and mammals into the state of New York. Trans. N. Amer. Wildl. Conf., 5:409-420.

Burt, W. H.
1946. The mammals of Michigan. Univ. Michigan Press, xv+288 pp.

Cahalane, V. H.
1964. A preliminary study of distribution and numbers of cougar, grizzly and wolf in North America. New York Zool. Soc., 12 pp.

Cahn, A. R.
1921. The mammals of Itasca County, Minnesota. Jour. Mamm., 2:68-74.

Carley, C. J., and H. McCarley
1976. An evaluation of red wolf (Canis rufus) hybridization in southeast Texas. Paper No. 153 at 56th Ann. Mtg. Amer. Soc. Mamm., Texas Tech Univ., Lubbock.

Carpenter, M.
1971. Some recent coyote records in Virginia. Virginia Wildl., 32(6):14-15.

Carson, H. S.
1962. Coyote, coy-dog, or dog. Maine Fish & Game, 4(1):4-7.

Catesby, M.
1771. The natural history of Carolina, Florida, and the Bahama Islands. London, 2 vols.

Chambers, R. E., P. N. Gaskin, R. A. Post, and S. A. Cameron
1974. The coyote. Conservationist, 29(2):5-7.

Chapman, F. M.
1894. Remarks on certain land mammals from Florida, with a list of the species known to occur in the state. Bull. Amer. Mus. Nat. Hist., 6:333-346.

Chiarelli, A. B.
1975. The chromosomes of the Canidae. In Fox (1975), pp. 40-53.

Churcher, C. S.
1959. Fossil Canis from the tar pits of La Brea, Peru. Science, 130:564-565.
1969a. The vertebrate fauna of Surprise, Mitchell and Island bluffs, near Medicine Hat, Alberta. Mid-Western Friends of the Pleistocene, 5 pp.
1969b. A fourth report on some Pleistocene localities in southern Alberta and Saskatchewan and their vertebrate fossil faunas. Geol. Surv. Canada, viii+410 pp.
1970. A fifth report on some Pleistocene localities in southern Alberta and Saskatchewan and their vertebrate fossil faunas. Geol. Surv. Canada, v+133 pp.

Clarke, C. H. D.
1970. Wolf management in Ontario. In Jorgensen, Faulkner, and Mech (1970), pp. 19-23.

Cleland, C. E.
1966. The prehistoric animal ecology and ethnozoology of the upper Great Lakes region. Univ. Michigan Anthropol. Papers, no. 29, x+294 pp.

Clutton-Brock, J., G. B. Corbett, and M. Hills
1976. A review of the family Canidae, with a classification by numerical methods. Bull. British Mus. (Nat. Hist.), Zool., 29:119-199.

Cockrum, E. L.
1952. Mammals of Kansas. Univ. Kansas Publ. Mus. Nat. Hist., 7:1-303.

Colbert, E. H.
1950. The fossil vertebrates. In Haury, E. W., The stratigraphy and archaeology of Ventana Cave Arizona, Univ. Arizona Press, Tucson, pp. 126-148.

Cole, F., and T. Deuel
1937. Rediscovering Illinois/archeological explorations in and around Fulton County. Univ. Chicago Press, xvi+295 pp.

Cook, R.
1952. The coy-dog: hybrid with a future? Jour. Hered., 43:71-73.

Cope, E. D.
1879. On the genera of Felidae and Canidae. Proc. Acad. Nat. Sci., Philadelphia, 1879, pp. 168-194.
1895. Extinct Bovidae, Canidae and Felidae from the Pleistocene of the plains. Jour. Acad. Nat. Sci., Philadelphia, ser. 2, 9:453-459.
1899. Vertebrate remains from Port Kennedy bone

deposit. Jour. Acad. Nat. Sci., Philadelphia, ser. 2, 11:193-267.

COPE, E. D., AND J. L. WORTMAN
1884. Post-Pliocene vertebrates of Indiana. Ann. Rept. State Geol. Indiana, 14:1-62.

CORGAN, J.
1976. Vertebrate fossils of Tennessee. Tennessee Div. Geol. Bull., no. 77, 100 pp.

CORY, C. B.
1912. The mammals of Illinois and Wisconsin. Field Mus. Nat. Hist. Publ., no. 153, Zool. Ser., 11:1-505.

COUES, E.
1873. The prairie wolf or coyote: *Canis latrans*. Amer. Nat., 7:385-389.

COWAN, I. M.
1954. The occurrence of the Pleistocene wolf, *Canis dirus*, in the Rocky Mountains of central Alberta. Canadian Field-Nat., 68:44.

CUNNINGHAM, V. D., AND R. D. DUNFORD
1970. Recent coyote record from Florida. Quart. Jour. Florida Acad. Sci., 33:279-280.

DALQUEST, W. W.
1961. Two species of bison contemporaneous in early Recent deposits in Texas. Southwestern Nat., 6:73-78.
1964. A new Pleistocene local fauna from Motley County, Texas. Trans. Kansas Acad. Sci., 67:499-505.
1965. New Pleistocene formation and local fauna from Hardeman County, Texas. Jour. Paleontol., 39:63-79.
1967. Mammals of the Pleistocene Slaton local fauna of Texas. Southwestern Nat., 12:1-30.
1968. Mammals of north-central Texas. Southwestern Nat., 13:13-21.
1975. Vertebrate fossils from the Blanco local fauna of Texas. Occas. Papers Mus. Texas Tech Univ., no. 30, 52 pp.

DALQUEST, W. W., E. ROTH, AND F. JUDD
1969. The mammal fauna of Schulze Cave, Edwards County, Texas. Bull. Florida State Mus., Biol. Sci., 13:205-276.

DAVIS, L. C.
1969. The biostratigraphy of Peccary Cave, Newton County, Arkansas. Proc. Arkansas Acad. Sci., 23:192-196.

DAVIS, W. B.
1966. The mammals of Texas. Texas Game and Fish Comm., Austin, 267 pp.

DE VOS, A.
1964. Range changes of mammals in the Great Lakes region. Amer. Midl. Nat., 71:210-231.

DEGERBØL, M.
1961. On a find of preboreal domestic dog (*Canis familiaris* L.) from Star Carr, Yorkshire, with remarks on other Mesolithic dogs. Proc. Prehist. Soc., 27:35-55.

DEL CAMPANA, D.
1913. I cani pliocenici di Toscana. Palaeontogr. Italica, 19:189-254.

DELLINGER, S. C., AND J. D. BLACK
1940. Notes on Arkansas mammals. Jour. Mamm., 21:187-191.

DICE, L. R.
1925. A survey of the mammals of Charlevoix County, Michigan, and vicinity. Occas. Papers Univ. Michigan Mus. Zool., no. 159, 33 pp.
1942. A family of dog-coyote hybrids. Jour. Mamm., 23:186-192.

DIXON, W. J.
1970. BMD biomedical computer programs. Univ. California Press, Berkeley, x+600 pp.

DOUGLASS, D. W.
1970. History and status of the wolf in Michigan. *In* Jorgensen, Faulkner, and Mech (1970), pp. 6-8.

DOWNS, T.
1958. Fossil vertebrates from Lago de Chapala, Jalisco, Mexico. Cong. Geol. Int., Mexico City, Sec. 7 — Paleontol., Taxon., and Evol., pp. 75-77.

DOWNS, T., H. HOWARD, T. CLEMENTS, AND G. A. SMITH
1959. Quaternary animals from Schuiling Cave in the Mojave Desert, California. Los Angeles Co. Mus. Contrib. Sci., no. 29, 21 pp.

DU BAR, J. R., AND G. CLOPINE
1962. Late Pleistocene deposits in the vicinity of Houston, Texas: a preliminary investigation. Trans. Gulf Coast Assoc. Geol. Societies, 11:83-108.

DUCK, L. G., AND J. B. FLETCHER
1945. A survey of the game and furbearing animals of Oklahoma. Oklahoma Game and Fish Comm., 144 pp.

EATON, G. F.
1923. Vertebrate fossils from the Miña Erupcion. Amer. Jour. Sci., ser. 5, 6:229-238.

ELDER, W. H., AND C. M. HAYDEN
1977. Use of discriminant function in taxonomic determination of canids from Missouri. Jour. Mamm., 58:17-24.

ELFTMAN, H. O.
1931. Pleistocene mammals of Fossil Lake, Oregon. Amer. Mus. Novit., no. 481, 21 pp.

ELLERMAN, J. R., AND T. C. S. MORRISON-SCOTT
1951. Checklist of Palaearctic and Indian mammals. British Mus., London, 810 pp.

FINE, M. D.
1964. An abnormal p2 in *Canis* cf. *C. latrans* from the Hagerman fauna of Idaho. Jour. Mamm., 45:483-485.

FISHER, O.
1841. Sketches of Texas in 1840. Walters and Weber, Springfield, Illinois, vii+64 pp.

FOX, M. W. (ed.)
1975. The wild canids/their systematics, behavioral ecology and evolution. Van Nostrand Reinhold, New York, xvi+508 pp.

FREEMAN, R. C.
1976. Coyote x dog hybridization and red wolf influence in the wild *Canis* of Oklahoma.

Unpublished M. A. thesis, Northeastern Oklahoma State Univ., vii+62 pp.

FRICK, C.
1930. Alaska's frozen fauna. Nat. Hist., 30(1): 71-80.

FUNKHOUSER, W. D.
1925. Wildlife in Kentucky. Kentucky Geol. Surv., Frankfort, 385 pp.

FURLONG, E. L.
1925. Notes on the occurrence of mammalian remains in the Pleistocene of Mexico, with description of a new species Capromeryx mexicana. Univ. California Publ. Geol. Sci., 15:137-152.

GABRIELSON, I. N.
1936. Report of the chief of the Bureau of Biological Survey. Washington, D.C.

GAGLIANO, S. M.
1967. Occupation sequence at Avery Island. Louisiana State Univ. Coastal Studies Ser., no. 22, xiii+110 pp.

GALBREATH, E. C.
1938. Post-glacial fossil vertebrates from east-central Illinois. Field Mus. Nat. Hist. Geol. Ser., 6:303-313.
1964. A dire wolf skeleton and Powder Mill Creek Cave, Missouri. Trans. Illinois State Acad. Sci., 57:224-242.

GALE, L. R., AND R. PIERCE
1954. Occurrence of the coyote in Kentucky. Jour. Mamm., 35:256-258.

GANIER, A. F.
1928. The wild life of Tennessee. Jour. Tennessee Acad. Sci., 3:10-22.

GARMAN, H.
1894. A preliminary list of the vertebrate animals of Kentucky. Bull. Essex Inst., 26: 1-63.

GAZIN, C. L.
1935. Annotated list of Pleistocene mammalia from American Falls, Idaho. Jour. Washington Acad. Sci., 25:297-302.
1942. The late Cenozoic vertebrate faunas from the San Pedro Valley, Arizona. Proc. U.S. Natl. Mus., 92:475-518.
1950. Annotated list of fossil mammalia associated with human remains at Melbourne, Florida. Jour. Washington Acad. Sci., 40:397-404.

GEIST, O. W.
1955. Vertebrate paleontological reconaissance of the Old Crow River area, Yukon Territory, Canada. Bull. Geol. Soc. Amer., 66:1702.

GEORGES, S.
1976. A range extension of the coyote in Quebec. Canadian Field-Nat., 90:78-79.

GETZ, L. L.
1960. Middle Pleistocene carnivores from southwestern Kansas. Jour. Mamm., 41:361-365.

GIDLEY, J. W.
1913. Preliminary report on a recently discovered Pleistocene cave deposit near Cumberland, Maryland. Proc. U. S. Natl. Mus., 46:93-102.

GIDLEY, J. W., AND C. L. GAZIN
1938. The Pleistocene vertebrate fauna from Cumberland Cave, Maryland. Bull. U. S. Natl. Mus., no. 171, vi+99 pp.

GIER, H. T.
1968. Coyotes in Kansas. Kansas State Univ. Agr. Exp. Sta. Bull., no. 393, 118 pp.

GIFFORD, C. L., AND R. WHITEBREAD
1951. Mammal survey of south central Pennsylvania. Pennsylvania Game Comm., 75 pp.

GILES, E.
1960. Multivariate analysis of Pleistocene and Recent coyotes (Canis latrans) from California. Univ. California Publ. Geol. Sci., 36:369-390.

GIPSON, P. S.
1972. The taxonomy, reproductive biology, food habits and range of wild Canis (Canidae) in Arkansas. Unpublished Ph.D. dissertation, Univ. Arkansas, viii+188 pp.
1976. Melanistic Canis in Arkansas. Southwestern Nat., 21:123-139.

GIPSON, P. S., I. K. GIPSON, AND J. A. SEALANDER
1975. Reproductive biology of wild Canis (Canidae) in Arkansas. Jour. Mamm., 56:605-612.

GIPSON, P. S., J. A. SEALANDER, AND J. E. DUNN
1974. The taxonomic status of wild Canis in Arkansas. Syst. Zool., 23:1-11.

GOERTZ, J. W., L. V. FITZGERALD, AND R. M. NOWAK
1975. The status of wild Canis in Louisiana. Amer. Midl. Nat., 93:215-218.

GOLDMAN, E. A.
1937. The wolves of North America. Jour. Mamm., 18:37-45.
1944. Classification of wolves. Part II in Young, S. P., and E. A. Goldman, The wolves of North America, Amer. Wildl. Inst., Washington, D.C., pp. 389-636.

GOLLEY, F. B.
1966. South Carolina mammals. Charleston Mus., xiv+181 pp.

GOODPASTER, W. W., AND D. F. HOFFMEISTER
1968. Notes on Ohioan mammals. Ohio Jour. Sci., 68:116-117.

GOODWIN, G. C.
1936. Big game animals in the northeastern United States. Jour. Mamm., 17:48-50.
1969. Mammals from the state of Oaxaca, Mexico, in the American Museum of Natural History. Bull. Amer. Mus. Nat. Hist., 141: 1-269.

GRAHAM, R.
1959. Additions to the Pleistocene fauna of Samwel Cave, California. I. Canis lupus and Canis latrans. Cave Studies, Univ. California Mus. Vert. Zool., 10:54-67.

GRAY, A. P.
1972. Mammalian hybrids. Commonwealth Agr. Bur., Slough, England, x+262 pp.

GREEN, M.
1948. A new species of dog from the lower Plio-

cene of California. Univ. California Publ. Bull. Dept. Geol. Sci., 28:81-90.

GRINNELL, J., J. S. DIXON, AND J. M. LINSDALE
1937. Fur-bearing mammals of California. Univ. California Press, Berkeley, 2:i-xiv+377-777.

GUILDAY, J. E.
1902. The Pleistocene local fauna of the Natural Chimneys, Augusta County, Virginia. Ann. Carnegie Mus., 36:87-122.
1969. Bone refuse from the Lamoka Lake site. In Ritchie (1969), pp. 54-59.
1971. Biological and archaeological analysis of bones from a 17th century Indian village (46 PU 31), Putnam County, West Virginia. West Virginia Geol. and Econ. Surv., Rept. Archeol. Invest., no. 4, vii+64 pp.
1977. The Clark's Cave bone deposit and the late Pleistocene paleoecology of the central Appalachian Mountains of Virginia. Bull. Carnegie Mus. Nat. Hist., no. 2, 87 pp.

GUILDAY, J. E., AND M. S. BENDER
1958. A recent fissure deposit in Bedford County, Pennsylvania. Ann. Carnegie Mus., 35:127-138.

GUILDAY, J. E., H. W. HAMILTON, AND A. D. McCRADY
1969. The Pleistocene vertebrate fauna of Robinson Cave, Overton County, Tennessee. Palaeovertebrata, 2:25-75.
1971. The Welsh Cave peccaries (*Platygonus*) and associated fauna, Kentucky Pleistocene. Ann. Carnegie Mus., 43:249-320.

GUILDAY, J. E., AND P. W. PARMALEE
1965. Animal remains from the Sheep Rock shelter (36 HU 1), Huntingdon County, Pennsylvania. Pennsylvania Archaeol., 35:34-49.

GUILDAY, J. E., P. W. PARMALEE, AND D. P. TANNER
1962. Aboriginal butchering techniques at the Eschelman site (36 La 12), Lancaster County, Pennsylvania. Pennsylvania Archaeol., 32:59-83.

GUILDAY, J. E., AND D. P. TANNER
1962. Animal remains from the Quaker State Rockshelter (36 Ve 27), Venango County, Pennsylvania. Pennsylvania Archaeol., 32:131-137.
1965. Vertebrate remains from the Mount Carbon site (46-Fa-7), Fayette County, West Virginia. West Virginia Archeol., 18:1-14.
1966. Vertebrate remains from the Fairchance Mound (46 Mr 13), Marshall County, West Virginia. West Virginia Archeol., 21:41-54.

GUT, H. J.
1939. Additions to the recorded Pleistocene mammals from Ocala, Florida. Proc. Florida Acad. Sci., 3:54-55.

GUT, H. J., AND C. E. RAY
1964. The Pleistocene vertebrate fauna of Reddick, Florida. Quart. Jour. Florida Acad. Sci., 26:315-328.

GUTHRIE, R. D.
1968. Paleoecology of the large-mammal community in interior Alaska during the late Pleistocene. Amer. Midl. Nat., 79:346-363.

HAAG, W. G.
1948. An osteometric analysis of some aboriginal dogs. Univ. Kentucky Repts. Anthropol., 7:107-264.

HAGER, M. W.
1972. A late Wisconsin-Recent vertebrate fauna from the Chimney Rock animal trap, Larimer County, Colorado. Univ. Wyoming Contrib. Geol., 11:63-71.

HALL, E. R.
1943. Cranial characters of a dog-coyote hybrid. Amer. Midl. Nat., 29:371-374.
1965. Names of species of North American mammals north of Mexico. Univ. Kansas Mus. Nat. Hist. Misc. Publ., no. 43, 16 pp.

HALL, E. R., AND K. R. KELSON
1952. Comments on the taxonomy and geographic distribution of some North American marsupials, insectivores and carnivores. Univ. Kansas Publ. Mus. Nat. Hist., 5:319-341.
1959. The mammals of North America. Ronald Press, New York, 2:i-viii+547-1083+1-79.

HALLORAN, A. F.
1958. Black red wolves. Oklahoma Wildl., 14(4):6-8.
1959. Shadow in the woods. Texas Game and Fish, 17(12):10-11.
1960. Black red wolf map. Texas Game and Fish, 18(3):25.
1961. The carnivores and ungulates of the Aransas National Wildlife Refuge, Texas. Southwestern Nat., 6:21-26.
1963. A melanistic coyote from Oklahoma. Southwestern Nat., 8:48-49.

HALLORAN, A. F., AND B. P. GLASS
1959. The carnivores and ungulates of the Wichita Mountains Wildlife Refuge, Oklahoma. Jour. Mamm., 40:360-370.

HAMILTON, W. J., JR.
1943. The mammals of eastern United States. Comstock Publ. Co., Ithaca, New York, 432 pp.

HAMNETT, W. L., AND D. C. THORNTON
1953. Tar heel wildlife. North Carolina Res. Comm., Raleigh.

HANDLAN, J. W.
1946. Hunter slays Monroe "wolf." West Virginia Conserv., 10(8):13, 23.

HANDLEY, C. O., JR., AND C. P. PATTON
1947. Wild mammals of Virginia. Virginia Comm. Game and Inland Fisheries, Richmond, vi+220 pp.

HARINGTON, C. R., AND F. V. CLULOW
1973. Pleistocene mammals from Gold Run Creek, Yukon Territory. Canadian Jour. Earth Sci., 10:697-759.

HARLAN, R.
1825. Fauna Americana. Philadelphia, x+318 pp.

HARPER, F.
1927. The mammals of the Okefinokee Swamp

region of Georgia. Proc. Boston Soc. Nat. Hist., 38:191-396.
1942. The name of the Florida wolf. Jour. Mamm., 23:339.

HARRINGTON, M. R.
1933. Gypsum Cave, Nevada. Southwest Mus. Papers, no. 8, ix+197 pp.

HARRIS, A. H.
1970. The Dry Cave mammalian fauna and late pluvial conditions in southeastern New Mexico. Texas Jour. Sci., 22:3-27.

HARRIS, A. H., AND J. J. FINDLEY
1964. Pleistocene-Recent fauna of the Isleta Caves, Bernalillo County, New Mexico. Amer. Jour. Sci., 262:114-120.

HARRISON, D. L.
1968. The mammals of Arabia. Ernest Benn, Ltd., London, 2:i-xiv+193-381.
1973. Some comparative features of the skulls of wolves (*Canis lupus* Linn.) and pariah dogs (*Canis familiaris* Linn.) from the Arabian Peninsula and neighbouring lands. Bonn Zool. Beitr., 24:185-191.

HAWKSLEY, O.
1963. The dire wolf in Missouri. Missouri Speleol., 5:63-72.
1965. Short-faced bear (Arctodus) fossils from Ozark caves. Bull. Natl. Speleol. Soc., 27:77-92.

HAWKSLEY, O., J. F. REYNOLDS, AND R. L. FOLEY
1973. Pleistocene vertebrate fauna of Bat Cave, Pulaski County, Missouri. Bull. Natl. Speleol. Soc., 35:61-87.

HAY, O. P.
1902. Bibliography and catalogue of the fossil vertebrata of North America. U. S. Geol. Surv., 868 pp.
1914. The Pleistocene mammals of Iowa. Rept. Iowa Geol. Surv., 23:1-662.
1917a. Vertebrata mostly from stratum number 3, at Vero, Florida, together with descriptions of new species. Ann. Rept. Florida Geol. Surv., 9:43-68.
1917b. On a collection of fossil vertebrates made by Dr. F. W. Cragin in the Equus beds of Kansas. Kansas Univ. Sci. Bull., 10:39-51.
1918. Quaternary vertebrates in southeastern Wisconsin. *In* Alden, W. C., The Quaternary geology of southeastern Wisconsin, U. S. Geol. Surv. Prof. Paper, no. 106, pp. 346-347.
1920. Descriptions of some Pleistocene vertebrates found in the United States. Proc. U. S. Natl. Mus., 58:83-146.
1921. Descriptions of species of Pleistocene vertebrata, types or specimens of which are preserved in the United States National Museum. Proc. U. S. Natl. Mus., 59:599-642.
1923. The Pleistocene of North America and its vertebrated animals from the states east of the Mississippi River and from the Canadian provinces east of longitude 95°. Carnegie Inst. Washington Publ., no. 322, viii+499 pp.
1924. The Pleistocene of the middle region of North America and its vertebrated animals. Carnegie Inst. Washington Publ., no. 322A, vii+385 pp.
1927. The Pleistocene of the western region of North America and its vertebrated animals. Carnegie Inst. Washington Publ., no. 322B, vii+346 pp.
1929-1930. Second bibliography and catalogue of the fossil vertebrata of North America. Carnegie Inst. Washington, 2 vols.

HENDRICKSON, J., AND W. L. ROBINSON
1975. Status of the wolf in Michigan, 1973. Amer. Midl. Nat., 94:226-232.

HESTER, J. J.
1960. Late Pleistocene extinction and radiocarbon dating. Amer. Antiquity, 26:58-77.

HIBBARD, C. W.
1938. An upper Pliocene fauna from Meade County, Kansas. Trans. Kansas Acad. Sci., 40:239-265.
1939. Notes on some mammals from the Pleistocene of Kansas. Trans. Kansas Acad. Sci., 42:463-479.
1941a. Mammals of the Rexroad fauna from the upper Pliocene of southwestern Kansas. Trans. Kansas Acad. Sci., 44:265-313.
1941b. Paleoecology and correlation of the Rexroad fauna from the upper Pliocene of southwestern Kansas, as indicated by the mammals. Univ. Kansas Sci. Bull., 27:79-104.
1949. Pleistocene stratigraphy and paleoecology of Meade County, Kansas. Contrib. Mus. Paleontol. Univ. Michigan, 7:63-90.
1953. Equus (Asinus) calobatus Troxell and associated vertebrates from the Pleistocene of Kansas. Trans. Kansas Acad. Sci., 56:111-126.
1955. Pleistocene vertebrates from the upper Becerra (Becerra Superior) formation, Valley of Tequixquiac, Mexico, with notes on other Pleistocene forms. Contrib. Mus. Paleontol. Univ. Michigan, 12:47-96.
1956. Vertebrate fossils from the Meade formation of southwestern Kansas. Papers Michigan Acad. Sci., Arts, and Letters, 41:145-203.
1958. Summary of North American Pleistocene mammalian local faunas. Papers Michigan Acad. Sci., Arts, and Letters, 43:3-32.
1970. Pleistocene mammalian local faunas from the Great Plains and central lowland provinces of the United States. *In* Dort, W., Jr., and J. K. Jones, Jr., Pleistocene and Recent environments of the central Great Plains, Univ. Kansas Dept. Geol., Spec. Publ. no. 3, pp. 395-433.

HIBBARD, C. W., AND W. W. DALQUEST
1966. Fossils from the Seymour formation of Knox and Baylor counties, Texas, and their

bearing on the late Kansan climate of that region. Contrib. Mus. Paleontol. Univ. Michigan, 21:1-66.

HIBBARD, C. W., J. C. FRYE, AND A. B. LEONARD
1944. Reconaissance of Pleistocene deposits in north-central Kansas. Kansas State Geol. Surv. Bull., no. 52, 28 pp.

HIBBARD, C. W., C. E. RAY, D. E. SAVAGE, D. W. TAYLOR, AND J. E. GUILDAY
1965. Quaternary mammals of North America. In Wright, H. E., and D. G. Frey (eds.), The Quaternary of the United States, Princeton Univ. Press, pp. 509-525.

HIBBARD, C. W., AND E. S. RIGGS
1949. Upper Pliocene vertebrates from Keefe Canyon, Meade County, Kansas. Bull. Geol. Soc. Amer., 60:829-860.

HIBBARD, C. W., AND D. W. TAYLOR
1960. Two late Pleistocene faunas from southwestern Kansas. Contrib. Mus. Paleontol. Univ. Michigan, 16:1-223.

HILDEBRAND, M.
1952a. An analysis of body proportions in the Canidae. Amer. Jour. Anat., 90:217-256.
1952b. The integument in Candidae. Jour. Mamm., 33:419-428.
1954. Comparative morphology of the body skeleton in Recent Canidae. Univ. California Publ. Zool., 55:399-470.

HILTON, H.
1977. More answers to Maine coyote questions. Maine Fish and Wildl., 19(1):2-4.

HIRSCHFELD, S. E.
1968. Vertebrate fauna of Nichol's Hammock, a natural trap. Quart. Jour. Florida Acad. Sci., 31:177-189.

HOFFMEISTER, D. F., AND W. W. GOODPASTER
1954. The mammals of the Huachuca Mountains, southeastern Arizona. Illinois Biol. Monogr., Univ. Illinois Press, no. 24, v+152 pp.

HOFFMEISTER, D. F., AND C. O. MOHR
1957. Fieldbook of Illinois mammals. Illinois Nat. Hist. Surv., xi+233 pp.

HOLLIMAN, D. C.
1963. The mammals of Alabama. Unpublished Ph.D. dissertation, Univ. Alabama, 504 pp.

HOOD, C. H., AND O. HAWKSLEY
1975. A Pleistocene fauna from Zoo Cave, Taney County, Missouri. Missouri Speleol., 15:1-42.

HOPKINS, M. L., R. BONNICHSEN, AND D. FORTSCH
1969. The stratigraphic position and faunal associates of Bison (Gigantobison) latifrons in southeastern Idaho, a progress report. Tebiwa, 12:1-8.

HOWARD, W. E.
1949. A means to distinguish skulls of coyotes and domestic dogs. Jour. Mamm., 30:169-171.

HOWELL, A. H.
1921. A biological survey of Alabama. N. Amer. Fauna, no. 45, 88 pp.

ILJIN, N. A.
1941. Wolf-dog genetics. Jour. Genetics, 42:359-414.

IMAIZUMI, Y.
1970a. Systematic status of the extinct Japanese wolf, Canis hodophilax. Jour. Mamm. Soc. Japan, 5:27-32.
1970b. Systematic status of the extinct Japanese wolf, Canis hodophilax, 2. Similarity relationship of hodophilax among the species of the genus Canis. Jour. Mamm. Soc. Japan, 5:62-66.

INTERNATIONAL COMMISSION ON ZOOLOGICAL NOMENCLATURE
1957. Opinion 447. Opinions and Declarations, 15 (part 12):211-224.

JACKSON, H. H. T.
1922. A coyote in Maryland. Jour. Mamm., 3:186-187.
1949. Two new coyotes from the United States. Proc. Biol. Soc. Washington, 62:31-32.
1951. Classification of the races of the coyote. Part II in Young, S. P., and H. H. T. Jackson, The Clever Coyote, Wildl. Mgmt. Inst., Washington, D.C., pp. 227-441.
1961. Mammals of Wisconsin. Univ. Wisconsin Press, Madison, xiv+504 pp.

JAKWAY, D. E.
1958. Pleistocene Lagomorpha and Rodentia from the San Josecito Cave, Neuvo Leon, Mexico. Trans. Kansas Acad. Sci., 61:313-327.

JENKINS, W.
1933. Wild life of Mississippi. Natchez, 155 pp.

JOHNSON, N. M., N. D. OPDYKE, AND E. H. LINDSAY
1975. Magnetic polarity stratigraphy of Pliocene-Pleistocene terrestrial deposits and vertebrate faunas, San Pedro Valley, Arizona. Bull. Geol. Soc. Amer., 86:5-12.

JOHNSTON, C. S.
1938. Preliminary report on the vertebrate type locality of Cita Canyon, and the description of an ancestral coyote. Amer. Jour. Sci., ser. 5, 35:383-390.

JOHNSTON, C. S., AND D. E. SAVAGE
1955. A survey of late Cenozoic vertebrate faunas of the Panhandle of Texas. Part I. Introduction, description of localities, preliminary faunal lists. Univ. California Publ. Geol. Sci., 31:27-50.

JOLICOEUR, P.
1959. Multivariate geographical variation in the wolf Canis lupus L. Evolution, 13:283-299.

JORGENSEN, S. E., C. E. FAULKNER, AND L. D. MECH (eds.)
1970. Proceedings of a symposium on wolf management in selected areas of North America. U. S. Bur. Sport Fisheries and Wildl., Twin Cities, Minnesota, iii+50 pp.

KEENER, J. M.
1970. History of the wolf in Wisconsin. In Jorgensen, Faulkner, and Mech (1970), pp. 4-5.

Kellogg, R.
1915. The mammals of Kansas with notes on their distribution, habits, life histories and economic importance. Unpublished M.A. thesis, Univ. Kansas, 310 pp.
1937. Annotated list of West Virginia mammals. Proc. U. S. Natl. Mus., 84:443-479.
1939. Annotated list of Tennessee mammals. Proc. U. S. Natl. Mus., 86:245-303.

Kennelly, J. J., and J. D. Roberts
1969. Fertility of coyote-dog hybrids. Jour. Mamm., 50:830-831.

Khan, E.
1970. Biostratigraphy and paleontology of a Sangamon deposit at Fort Qu'Appelle, Saskatchewan. Natl. Mus. Canada Publ. Paleontol., no. 5, viii+82 pp.

Klein, J.
1971. The ferungulates of the Inglis IA local fauna, early Pleistocene of Florida. Unpublished M.S. thesis, Univ. Florida, 115 p.

Kolenosky, G. B.
1971. Hybridization between wolf and coyote. Jour. Mamm., 52:446-449.

Kolenosky, G. B., and R. O. Standfield
1975. Morphological and ecological variation among gray wolves (*Canis lupus*) of Ontario, Canada. In Fox (1975), pp. 62-72.

Kraglievich, J. L.
1952. Un canido del Eocuartario de Mar del Plata y sus relaciones con otras formas Brasilenas y Norteamericanas. Rev. Mus. Mar del Plata, 1:53-70.

Kraglievich, L.
1928. Contribucion al conocimiento de los grandes canidos extinguidos de Sud America. An. Soc. Cien. Argentina, 106:332-342.

Krefting, L. W.
1969. The rise and fall of the coyote on Isle Royale. Naturalist, 20(4):24-31.

Kurten, B.
1963. Notes on some Pleistocene mammal migrations from the Palaearctic to the Nearctic. Eiszeitalter u. Gegenwart, 14:96-103.
1967. Prariewolf und sabelzahntiger aus dem Pleistozan des Valsequillo, Mexiko. Quartar, 18:173-178.
1968. Pleistocene mammals of Europe. Aldine, Chicago, viii+317 pp.
1974. A history of coyote-like dogs (Canidae, Mammalia). Acta Zool. Fennica, no. 140, 38 pp.

Kurten, B., and E. Anderson
1972. The sediments and fauna of Jaguar Cave. II—The fauna. Tebiwa, 15:21-45.

Langdon, F. W.
1881. The mammalia of the vicinity of Cincinnati—a list of species with notes. Jour. Cincinnati Soc. Nat. Hist., 1:297-313.

Langguth, A.
1975. Ecology and evolution in the South American canids. In Fox (1975), pp. 192-206.

Lantz, D. E.
1905. A list of Kansas mammals. Trans. Kansas Acad. Sci., 19:171-178.

Lawrence, B.
1966. Early domestic dogs. Zeit. f. Saugetierkunde, 32:44-59.
1968. Antiquity of large dogs in North America. Tebiwa, 11:43-49.

Lawrence, B., and W. H. Bossert
1967. Multiple character analysis of *Canis lupus*, *latrans*, and *familiaris*, with a discussion of the relationships of *Canis niger*. Amer. Zool., 7:223-232.
1969. The cranial evidence for hybridization in New England Canis. Breviora, no. 330, 13 pp.
1975. Relationships of North American *Canis* shown by a multiple character analysis of selected populations. In Fox (1975), pp. 73-86.

Leidy, J.
1854. Note on some fossil bones discovered by Mr. Francis A. Lincke in the banks of the Ohio River, Indiana. Proc. Acad. Nat. Sci., Philadelphia, 7:199-201.
1856. Description of some remains of extinct mammalia. Jour. Acad. Nat. Sci., Philadelphia, ser. 2, 3:166-171.
1858. Notice of remains of extinct vertebrata, from the Valley of the Niobrara River. Proc. Acad. Nat. Sci., Philadelphia, 1858, pp. 20-29.
1869. The extinct mammalian fauna of Dakota and Nebraska. Jour. Acad. Nat. Sci., Philadelphia, ser. 2, 7:1-472.
1873. Contributions to the extinct vertebrate fauna of the western territories. Rept. U. S. Geol. Surv., 358 pp.
1889. Notice and description of fossils in caves and crevices of the limestone rocks of Pennsylvania. Ann. Rept. Geol. Surv. Pennsylvania, 1887, pp. 1-20.

Leopold, A. S., and E. R. Hall
1945. Some mammals of Ozark County, Missouri. Jour. Mamm., 26:142-145.

Linhart, S. B., and F. F. Knowlton
1967. Determining age of coyotes by tooth cementum layers. Jour. Wildl. Mgmt., 31:362-365.

Linzey, D. W.
1971. Animal harvested in south Alabama probably coyote-red wolf hybrid. Alabama Conserv., 41(6):6-7.

Linzey, D. W., and A. V. Linzey
1968. Mammals of the Great Smoky Mountains National Park. Jour. Elisha Mitchell Sci. Soc., 84:384-414.

Loomis, F. B., and D. B. Young
1912. On the shell heaps of Maine. Amer. Jour. Sci., ser. 4, 34:17-42.

Lowery, G. H., Jr.
1943. Check-list of the mammals of Louisiana

and adjacent waters. Occas. Papers Louisiana State Univ. Mus. Zool., 13:213-257.

LUNDELIUS, E. L., JR.
1960. *Mylohyus nasutus* long-nosed peccary of the Texas Pleistocene. Bull. Texas Mem. Mus., no. 1, 40 pp.
1962. Late Pleistocene vertebrate fauna from San Patricio County, Texas. Geol. Soc. Amer. Spec. Paper, no. 68, p. 222.
1967. Late-Pleistocene and Holocene faunal history of central Texas. *In* Martin and Wright (1967), pp. 287-319.
1972. Fossil vertebrates from the late Pleistocene Ingleside fauna, San Patricio County, Texas. Univ. Texas, Bur. Econ. Geol. Rept. Invest., no. 77, 74 pp.

LYON, M. W., JR.
1936. Mammals of Indiana. Amer. Midl. Nat., 17:1-384.

MACPHERSON, A. H.
1965. The origin of diversity in mammals of the Canadian Arctic tundra. Syst. Zool., 14:153-173.

MAINE DEPARTMENT OF INLAND FISHERIES AND WILDLIFE
1976. Coyote. Maine Fish and Wildl., 18(4):S-19.

MALDONADO-KOERDELL, M.
1955. Sobre un craneo de *Aenocyon dirus* (Leidy) del Pleistoceno superior de Tequixquiac, Mexico. An. Inst. Nac. Antro. Hist. Mexico, 7:51-58.

MANNING, T. H., AND A. H. MACPHERSON
1958. The mammals of Banks Island. Arctic Inst. N. Amer. Tech. Paper, no. 2, 74 pp.

MANVILLE, R. H., AND W. C. STURTEVANT
1966. Early specimens of the eastern wolf, *Canis lupus lycaon*. Chesapeake Sci., 7:218-219.

MARCUS, L. F.
1960. A census of the abundant large Pleistocene mammals from Rancho La Brea. Los Angeles Co. Mus. Contrib. Sci., no. 38, 11 pp.

MARTIN, L. D.
1972. The microtine rodents of the Mullen assemblage from the Pleistocene of north central Nebraska. Bull. Univ. Nebraska State Mus., 9:173-182.

MARTIN, P. S., AND H. E. WRIGHT (eds.)
1967. Pleistocene extinctions. Yale Univ. Press, New Haven, x+453 pp.

MARTIN, R. A.
1974. Fossil mammals from the Coleman IIA fauna, Sumter County. *In* Webb (1974a), pp. 35-99.

MARTIN, R. A., AND S. D. WEBB
1974. Late Pleistocene mammals from the Devil's Den fauna, Levy County. *In* Webb (1974a), pp. 114-145.

MARVINNEY, S.
1976. Will the predators return? Conservationist, 31(1):iv-v.

MATTHEW, W. D.
1902. List of the Pleistocene fauna from Hay Springs, Nebraska. Bull. Amer. Mus. Nat. Hist., 16:317-322.
1916. The grim wolf of the tar pits. Amer. Mus. Jour., 16(1):45-47.
1918. Contributions to the Snake Creek fauna with notes upon the Pleistocene of western Nebraska American Museum expedition of 1916. Bull. Amer. Mus. Nat. Hist., 38:183-229.
1930. The phylogeny of dogs. Jour. Mamm., 11:117-138.

MAWBY, J. E.
1967. Fossil vertebrates of the Tule Springs site, Nevada. Nevada State Mus. Anthropol. Papers, 13:106-128.

MAYR, E.
1963. Animal species and evolution. Harvard Univ. Press, Cambridge, xiv+797 pp.

MCCARLEY, H.
1959. The mammals of eastern Texas. Texas Jour. Sci., 11:385-426.
1962. The taxonomic status of wild *Canis* (Canidae) in the south central United States. Southwestern Nat., 7:227-235.

MCDONALD, H. G., AND E. ANDERSON
1975. A late Pleistocene vertebrate fauna from southeastern Idaho. Tebiwa, 18:19-37.

MCGREW, P. O.
1944. An early Pleistocene (Blancan) fauna from Nebraska. Field Mus. Nat. Hist. Geol. Ser., 9:33-66.

MCKNIGHT, T.
1964. Feral livestock in Anglo-America. Univ. California Publ. Geogr., 16:1-78.

MEADE, G. E.
1945. The Blanco fauna. Univ. Texas Publ., 4401:509-556.

MECH, L. D.
1959. The coyote comes east. Frontiers, 23:117-119, 126.
1961. Exit timber wolf, enter coyote. Animal Kingdom, 64(3):89-92.
1966. The wolves of Isle Royale. Fauna Natl. Parks, U.S., Fauna Ser., no. 7, xiv+210 pp.
1970. The wolf: the ecology and behavior of an endangered species. Natural History Press, Garden City, New York, xx+384 pp.
1977. A recovery plan for the eastern timber wolf. Natl. Parks and Conserv. Mag., 51(1):17-21.

MECH, L. D., AND L. D. FRENZEL, JR.
1971. The possible occurrence of the Great Plains wolf in northeastern Minnesota. *In* Mech, L. D., and L. D. Frenzel, Jr. (eds.), Ecological studies of the timber wolf in northeastern Minnesota, N. Cent. Forest Exp. Sta., St. Paul, pp. 60-62.

MEHL, M. G.
1962. Missouri's ice age mammals. Missouri Div. Geol. and Water Res., Ed. Ser., no. 1, xi+104 pp.

MENGEL, R. M.
1971. A study of dog-coyote hybrids and implications concerning hybridization in *Canis*. Jour. Mamm., 52:316-336.

MERRIAM, C. H.
1897. Revision of the coyotes or prairie wolves, with descriptions of new forms. Proc. Biol. Soc. Washington, 11:19-33.

MERRIAM, J. C.
1903. The Pliocene and Quaternary Canidae of the Great Valley of California. Univ. California Publ. Bull. Dept. Geol., 3:277-290.
1906. Recent discoveries of Quaternary mammals in southern California. Science, n.s., 24:248-250.
1910. New Mammalia from Rancho La Brea. Univ. California Publ. Bull. Dept. Geol., 5:391-395.
1911. Tertiary mammal beds of Virgin Valley and Thousand Creek in northwestern Nevada. Univ. California Publ. Bull. Dept. Geol., 6:199-304.
1912. The fauna of Rancho La Brea. Part II. Canidae. Mem. Univ. California, 1:217-272.
1918. Note on the systematic position of the wolves of the Canis dirus group. Univ. California Publ. Bull. Dept. Geol., 10:531-533.

MERRIAM, J. C., AND C. STOCK
1921. Occurrence of Pleistocene vertebrates in an asphalt deposit near McKittrick, California. Science, n.s., 54:566-567.

MILLER, A. M.
1922. Licks and caves of the lower Ohio Valley as repositories of mammalian remains, including those of man. Bull. Geol. Soc. Amer., 33:156-159.

MILLER, G. J.
1968. On the age distribution of *Smilodon californicus* Bovard from Rancho La Brea. Los Angeles Co. Mus. Contrib. Sci., no. 131, 17 pp.

MILLER, G. S., JR.
1899. Preliminary list of New York mammals. Bull. New York State Mus., 6:271-390.
1912a. The names of two North American wolves. Proc. Biol. Soc. Washington, 25:95.
1912b. The names of the large wolves of northern and western North America. Smithsonian Misc. Coll., 59(15):1-5.
1912c. Catalogue of the mammals of western Europe (Europe exclusive of Russia). British Mus., London, xv+1019 pp.

MILLER, M. E., G. C. CHRISTENSEN, AND H. E. EVANS
1964. Anatomy of the dog. W. B. Saunders Co., Philadelphia, xii+941 pp.

MILLER, W. E.
1971. Pleistocene vertebrates of the Los Angeles Basin and vicinity (exclusive of Rancho La Brea). Bull. Los Angeles Co. Mus. Nat. Hist., Sci.: no. 10, 124 pp.
1976. Late Pleistocene vertebrates of the Silver Creek local fauna from north central Utah. Great Basin Nat., 36:387-424.

MIVART, ST. G.
1890. Monograph of the Canidae. London, 216 pp.

MOOSER, O., AND W. W. DALQUEST
1975. Pleistocene mammals from Aguascalientes, central Mexico. Jour. Mamm., 56:781-820.

MORRIS, R. F.
1948. The land mammals of New Brunswick. Jour. Mamm., 29:165-176.

MORRISON, J. D.
1970. The Eddy Bluff shelter of Beaver Reservoir of northwest Arkansas. Proc. Arkansas Acad. Sci., 24:85-91.

MORRISON, R. B.
1964. Lake Lohontan: geology of southern Carson Desert, Nevada. U. S. Geol. Surv. Prof. Paper, no. 401, v+156 pp.

MUMFORD, R. E.
1969. Distribution of the mammals of Indiana. Indiana Acad. Sci. Monogr., no. 1, vii+114 pp.

MURPHY, J. L.
1968. The Hobson site: a Fort Ancient component near Middleport, Meigs County, Ohio. Kirtlandia, no. 4, 14 pp.

NEGUS, N. C.
1948. A coyote, *Canis latrans*, from Preble County, Ohio. Jour. Mamm., 29:295.

NESBITT, W. H.
1975. Ecology of a feral dog pack on a wildlife refuge. *In* Fox (1975), pp. 391-396.

NIGRA, J. O., AND J. F. LANCE
1947. A statistical study of the metapodials of the dire wolf group from the Pleistocene of Rancho La Brea. Bull. S. California Acad. Sci., 46:26-34.

NOWAK, R. M.
1967. The red wolf in Louisiana. Defenders of Wildl. News, 42:60-70.
1970. Report on the red wolf. Defenders of Wildl. News, 45:82-94.
1971. Louisiana protects wolf, cougar, and all birds of prey. Defenders of Wildl. News, 46:278.
1972. The mysterious wolf of the south. Nat. Hist., 81(1):50-53, 74-77.
1973. North American Quaternary Canis. Unpublished Ph.D. dissertation, Univ. Kansas, 380 pp.
1974. Red wolf: our most endangered mammal. Natl. Parks and Conserv. Mag., 48(8):9-12.

OLSEN, S. J., AND J. W. OLSEN
1977. The Chinese wolf, ancestor of New World dogs. Science, 197:533-535.

OLSON, E. C.
1940. A late Pleistocene fauna from Herculaneum, Missouri. Jour. Geol., 48:32-57.

OSGOOD, W. H.
1934. The genera and subgenera of South American canids. Jour. Mamm., 15:45-50.

OZOGA, J. J., AND E. M. HARGER
1966. Occurrence of albino and melanistic coyotes in Michigan. Jour. Mamm., 47:339-340.

PACKARD, A. S.
1885. Origin of the American varieties of the dog. Zoologist, ser. 3, 9:367-372.

PACKARD, E. L.
1950. A large wolf from the Pleistocene of Willamette Valley, Oregon. Geol. News Letter, Geol. Soc. Oregon Country, 16:89-90.

PARADISO, J. L.
1965. Recent records of red wolves from the Gulf Coast of Texas. Southwestern Nat., 10:318-319.
1966. Recent records of coyotes, Canis latrans, from the southeastern United States. Southwestern Nat., 11:500-501.
1968. Canids recently collected in east Texas, with comments on the taxonomy of the red wolf. Amer. Midl. Nat., 80:529-534.
1969. Mammals of Maryland. N. Amer. Fauna, no. 66, iv+193 pp.

PARADISO, J. L., AND R. M. NOWAK
1972a. A report on the taxonomic status and distribution of the red wolf. U. S. Bur. Sport Fisheries and Wildl. Spec. Sci. Rept.—Wildl., no. 145, ii+36 pp.
1972b. Canis rufus. Mammalian Species, Amer. Soc. Mamm., no. 22, 4 pp.
1973. New data on the red wolf in Alabama. Jour. Mamm., 54:506-509.

PARADISO, J. L., AND D. SCHIERBAUM
1969. Recent wolf record from New York. Jour. Mamm., 50:384-385.

PARMALEE, P. W.
1957. Vertebrate remains from the Cahokia site, Illinois. Trans. Illinois State Acad. Sci., 50:235-242.
1959a. Use of mammalian skulls and mandibles by prehistoric Indians of Illinois. Trans. Illinois State Acad. Sci., 52:85-95.
1959b. Animal remains from the Raddatz rockshelter, Sk 5, Wisconsin. Wisconsin Archeol., 40:83-90.
1959c. Animal remains from the Banks site, Crittenden County, Arkansas. Tennessee Archaeol. Soc. Misc. Paper, no. 5, 8 pp.
1962a. Additional faunal records from the Kingston Lake site, Illinois. Trans. Illinois State Acad. Sci., 55:6-12.
1962b. The faunal complex of the Fisher site, Illinois. Amer. Midl. Nat., 68:399-408.
1963. Vertebrate remains from the Bell site, Winnebago County, Wisconsin. Wisconsin Archeol., 44:58-69.
1965. The food economy of Archaic and Woodland peoples at the Tick Creek Cave site, Missouri. Missouri Archaeol., 27:1-34.
1972. Vertebrate remains from the Fifield site, Porter County, Indiana. Indiana Hist. Soc. Prehist. Res. Ser., 5:202-205.

PARMALEE, P. W., AND R. D. OESCH
1972. Pleistocene and Recent faunas from the Brynjulfson Caves, Missouri. Illinois State Mus. Rept. Invest., no. 25, vii+52 pp.

PARMALEE, P. W., AND O. C. SHANE, III
1970. The Blain site vertebrate fauna. In Prufer, O. H., and O. C. Shane, III, Blain Village and the Fort Ancient tradition in Ohio, Kent State Univ. Press, pp. 185-206.

PARMALEE, P. W., AND D. STEPHENS
1972. A wolf mask and other carnivore skull artifacts from the Palestine site, Illinois. Pennsylvania Archaeol., 42:71-74.

PATTERSON, B.
1932. Upper molars of Canis armbrusteri Gidley from Cumberland Cave, Maryland. Amer. Jour. Sci., ser. 5, 23:334-336.

PATTON, T. H.
1963. Fossil vertebrates from Miller's Cave, Llano County, Texas. Bull. Texas Mem. Mus., 7:1-41.

PAUL, J. R.
1970. Coyotes and kin. Explorer, 12(1):23-25.

PETERSON, O. A.
1926. The fossils of the Frankstown Cave, Blair County, Pennsylvania. Ann. Carnegie Mus., 16:249-315.

PETERSON, R. L.
1946. Recent and Pleistocene mammalian fauna of Brazos County, Texas. Jour. Mamm., 27:162-169.
1957. Changes in the mammalian fauna of Ontario. In Urquhart, F. A. (ed.), Changes in the fauna of Ontario, Contrib. Royal Ontario Mus., pp. 43-58.
1966. The mammals of eastern Canada. Oxford Univ. Press, Toronto, xxxii+465 pp.

PEWE, T. L., AND D. M. HOPKINS
1967. Mammal remains of pre-Wisconsin age in Alaska. In Hopkins, D. M. (ed.), The Bering Land Bridge, Stanford Univ. Press, pp. 266-270.

PIMLOTT, D. H., AND P. W. JOSLIN
1968. The status and distribution of the red wolf. Trans. N. Amer. Wildl. and Nat. Res. Conf., 33:373-389.

POCOCK, R. I.
1935. The races of Canis lupus. Proc. Zool. Soc. London, 1935, pt. 3, pp. 617-686.

PRINGLE, L. P.
1960. Notes on coyotes in southern New England. Jour. Mamm., 41:278.
1963. Coyotes in New England. Massachusetts Audubon, 48(2):52-54.

QUACKENBUSH, L. S.
1909. Notes on Alaskan mammoth expedition of 1907 and 1908. Bull. Amer. Mus. Nat. Hist., 26:87-130.

QUINN, J. H.
1972. Extinct mammals in Arkansas and related C^{14} dates circa 3000 years ago. Int. Geol. Congr., 24(sect. 12):89-96.

RAND, A. L.
1945. Mammals of the Ottawa district. Canadian Field-Nat., 59:111-132.

Rao, C. R.
1952. Advanced statistical methods in biometric research. Wiley, New York, 390 pp.

Ray, C. E.
1958. Additions to the Pliestocene mammalian fauna from Melbourne, Florida. Bull. Mus. Comp. Zool., 119:421-451.
1967. Pleistocene mammals from Ladds, Bartow County, Georgia. Georgia Acad. Sci. Bull., 25:120-150.

Redington, P. G.
1931. Report of the chief of the Bureau of Biological Survey. Washington, D.C.

Reed, C. A.
1961. Osteological evidences for prehistoric domestication in southwestern Asia. Sonderdruck Zeit. Tierzuchtung und Zuchtungsbiologie, 76:31-38.

Reynolds, H. S.
1909. A monograph of the British Pleistocene mammalia, vol. 2, part 3, pp. 1-28, The Canidae. London Palaeontographical Soc.

Rhoads, S. N.
1903. The mammals of Pennsylvania and New Jersey. Philadelphia, 266 pp.

Richardson, J.
1829. Fauna Boreali-Americana. John Murray, London, xvi+300 pp.

Richens, V. B., and R. D. Hugie
1974. Distribution, taxonomic status, and characteristics of coyotes in Maine. Jour. Wildl. Mgmt., 38:447-454.

Richmond, N. D., and H. R. Rosland
1949. Mammal survey of northwestern Pennsylvania. Pennsylvania Game Comm., 87 pp.

Rinker, G. C., and C. W. Hibbard
1952. A new beaver and associated vertebrates, from the Pleistocene of Oklahoma. Jour. Mamm., 33:98-101.

Ritchie, W. A.
1969. The archaeology of New York State. Natural History Press, Garden City, New York, xxxiv+357 pp.

Roemer, F.
1849. Texas. 1935 reprint by Standard Printing Co., San Antonio, xii+301 pp.

Roth, E. L.
1972. Late Pleistocene mammals from Klein Cave, Kerr County, Texas. Texas Jour. Sci., 24: 75-84.

Russell, D. N., and J. H. Shaw
1971a. Notes on the red wolf (*Canis rufus*) in the coastal marshes and prairies of eastern Texas. Texas Parks and Wildl. Dept., 5 pp.
1971b. Distribution and relative density of the red wolf in Texas. Texas Parks and Wildl. Dept., 11 pp.
1972. The red wolf—situation critical. Texas Parks and Wildl., 30(3):12-15.

Russell, R. J.
1960. Pleistocene pocket gophers from San Josecito Cave, Nuevo Leon, Mexico. Univ. Kansas Publ. Mus. Nat. Hist., 9:539-548.

St. Amant, L. S.
1959. Louisiana wildlife inventory and management plan. Louisiana Wild Life and Fisheries Comm., xx+329 pp.

Sampson, F. W.
1961. Missouri's vanishing wolves. Missouri Conserv., 22(6):5-7.

Saunders, J. J.
1977. Late Pleistocene vertebrates of the western Ozark Highland, Missouri. Illinois State Mus. Rept. Invest., no. 33, x+118 pp.

Savage, D. E.
1951. Late Cenozoic vertebrates of the San Francisco Bay region. Univ. California Publ. Bull. Dept. Geol. Sci., 28:215-314.

Schorger, A. W.
1942. Extinct and endangered mammals and birds of the upper Great Lakes region. Trans. Wisconsin Acad. Sci., Arts and Letters, 34: 23-44.

Schultz, C. B.
1934. The Pleistocene mammals of Nebraska. Bull. Nebraska State Mus., 1:357-393.

Schultz, C. B., and E. B. Howard
1935. The fauna of Burnet Cave, Guadalupe Mountains, New Mexico. Proc. Acad. Nat. Sci., Philadelphia, 87:273-298.

Schultz, C. B., and L. D. Martin
1970. Quaternary mammalian sequence in the Great Plains. In Dort, W., Jr., and J. K. Jones, Jr., Pleistocene and Recent environments of the central Great Plains, Univ. Kansas Dept. Geol. Spec. Publ., no. 3, pp. 341-353.

Schultz, C. B., L. D. Martin, and L. G. Tanner
1970. Mammalian distribution in the Great Plains and adjacent areas from 14,000 to 9,000 years ago. Abstr., Amer. Quat. Assoc. Mtg., 1:119-120.

Schultz, C. B., and T. M. Stout
1945. Pleistocene loess deposits of Nebraska. Amer. Jour. Sci., 243:231-244.
1948. Pleistocene mammals and terraces in the Great Plains. Bull. Geol. Soc. Amer., 59: 533-588.

Schultz, C. B., and L. G. Tanner
1957. Medial Pleistocene fossil vertebrate localities in Nebraska. Bull. Univ. Nebraska State Mus., 4:59-81.

Schultz, G. E.
1969. Geology and paleontology of a late Pleistocene basin in southwest Kansas. Geol. Soc. Amer. Spec. Paper, no. 105, viii+85 pp.

Schultz, J. R.
1938a. Early Pleistocene mammal fauna from the vicinity of Grand View, Ada and Owyhee counties, Idaho. Proc. Geol. Soc. Amer., 1937, p. 297.
1938b. A late Quaternary mammal fauna from the tar seeps of McKittrick, California. Carnegie Inst. Washington Publ., 487:111-215.

Schultz, V.
1955. Status of the coyote and related forms in

Tennessee. Jour. Tennessee Acad. Sci., 30: 44-46.

SCHULTZ, V., et al.
1954. Statewide wildlife survey of Tennessee. Tennessee Game and Fish Comm., Nashville, 506 pp.

SCOTT, J. P.
1968. Evolution and domestication of the dog. Evol. Biol., 2:243-275.

SCOTT, J. P., AND J. L. FULLER
1965. Genetics and the social behavior of the dog. Univ. Chicago Press, xviii+468 pp.

SCOTT, W. B.
1937. A history of land mammals in the Western Hemisphere. Amer. Phil. Soc., New York, xiv+786 pp.

SEAL, H.
1964. Multivariate statistical analysis for biologists. Wiley, New York, 207 pp

SEALANDER, J. A., JR.
1956. A provisional check-list and key to the mammals of Arkansas (with annotations). Amer. Midl. Nat., 56:257-296.

SELLARDS, E. H.
1916. Human remains and associated fossils from the Pleistocene of Florida. Ann. Rept. Florida Geol. Surv., 8:123-160.
1940. Pleistocene artifacts and associated fossils from Bee County, Texas. Bull. Geol. Soc. Amer., 51:1627-1658.

SEMKEN, H. A.
1961. Fossil vertebrates from Longhorn Cavern, Burnet County, Texas. Texas Jour. Sci., 13:290-310.

SERRES, M.
1835. On the distinctive characters of the dog, the wolf, and the fox, as supplied by the skeleton. Edinburgh New Philos. Jour., 19: 244-253.

SETON, E. T.
1929. Lives of game animals. Doubleday, Doran & Co., Garden City, New York, 1(part 2): 339-640.

SEVERINGHAUS, C. W.
1974a. Notes on the history of wild canids in New York. New York Fish and Game Jour., 21: 117-125.
1974b. The coyote moves east. Conservationist, 29(2):8, 36.

SHAKESPEAR, S.
1975. An osteometric description of the Pleistocene dire wolf, *Canis dirus*, from the Maricopa Brea of California. Proc. Utah Acad. Sci. Arts Letters, 52:77-78.

SHAW, J. H.
1975. Ecology, behavior, and systematics of the red wolf (*Canis rufus*). Unpublished Ph.D. dissertation, Yale Univ., v+99+xi pp.

SHAY, C. T.
1963. A preliminary report on the Itasca bison site. Proc. Minnesota Acad. Sci., 31:24-27.

SHERMAN, H. B.
1937. List of the Recent wild land mammals of Florida. Proc. Florida Acad. Sci., 1:102-128.

SHIRAS, G., III
1921. The wildlife of Lake Superior, past and present. Natl. Geogr., 40:113-204.

SHOEMAKER, H. W.
1917. Extinct Pennsylvania animals. Altoona Tribune, Altoona, Pennsylvania, 201 pp.

SHOTWELL, J. A.
1956. Hemphillian mammalian assemblage from northeastern Oregon. Bull. Geol. Soc. Amer., 67:717-738.
1970. Pliocene mammals of southeast Oregon and adjacent Idaho. Bull. Mus. Nat. Hist. Univ. Oregon, 17:1-103.

SILVER, H.
1957. A history of New Hampshire game and furbearers. New Hampshire Fish & Game Dept. Surv. Rept., no. 6, xiv+466 pp.

SILVER, H., AND W. T. SILVER
1969. Growth and behavior of the coyote-like canid of northern New England with observations on canid hybrids. Wildl. Monogr., no. 17, 41 pp.

SIMPSON, G. G.
1928. Pleistocene mammals from a cave in Citrus County, Florida. Amer. Mus. Novit., no. 328, 16 pp.
1929a. Pleistocene mammalian fauna of the Seminole Field, Pinellas County, Florida. Bull. Amer. Mus. Nat. Hist., 56:561-599.
1929b. The extinct land mammals of Florida. Ann. Rept. Florida Geol. Surv., 20:229-279.
1941. Large Pleistocene felines of North America. Amer. Mus. Novit., no. 1136, 27 pp.
1945. The principles of classification and a classification of mammals. Bull. Amer. Mus. Nat. Hist., 85:v-xvi+1-350.
1949. A fossil deposit in a cave in St. Louis. Amer. Mus. Novit., no. 1408, 46 pp.

SINCLAIR, W. J.
1904. The exploration of the Potter Creek Cave. Univ. California Publ. Amer. Archaeol. and Ethnol., 2:1-27.

SKAGGS, O.
1946. A study of the dog skeletons from Indian Knoll with special reference to the coyote as a progenitor. *In* Webb, W. S., Indian Knoll, Univ. Kentucky Rept. Anthropol. and Archaeol., 4:341-355.

SKINNER, M. F.
1942. The fauna of Papago Springs Cave, Arizona. Bull. Amer. Mus. Nat. Hist., 80:143-220.

SKINNER, M. F., AND C. W. HIBBARD
1972. Early Pleistocene preglacial and glacial rocks and faunas of north-central Nebraska. Bull. Amer. Mus. Nat. Hist., 148:1-148.

SLAUGHTER, B. H.
1961. A new coyote in the late Pleistocene of Texas. Jour. Mamm., 42:503-509.
1966a. The Moore pit local fauna; Pleistocene of Texas. Jour. Paleontol., 40:78-91.

1966b. *Platygonus compressus* and associated fauna from the Laubach Cave of Texas. Amer. Midl. Nat., 75:475-494.

SLAUGHTER, B. H., W. W. CROOK, JR., R. K. HARRIS, D. C. ALLEN, AND M. SEIFERT
1962. The Hill-Shuler local faunas of the upper Trinity River, Dallas and Denton counties, Texas. Univ. Texas, Bur. Econ. Geol. Rept. Invest., no. 48, viii+75 pp.

SLAUGHTER, B. H., AND B. R. HOOVER
1963. Sulphur River formation and the Pleistocene mammals of the Ben Franklin local fauna. Jour. Grad. Res. Center, S. Meth. Univ., 31:132-148.

SLAUGHTER, B. H., AND R. RITCHIE
1963. Pleistocene mammals of the Clear Creek local fauna, Denton County, Texas. Jour. Grad. Res. Center, S. Meth. Univ., 31:117-131.

SMITS, L.
1963. King of the wild. Michigan Conserv., 32(1):45-50.

SNYDER, L. L.
1938. A faunal investigation of western Rainy River District, Ontario. Trans. Royal Canadian Inst., Toronto, 22:157-180.

STANDFIELD, R.
1970. Some considerations on the taxonomy of wolves in Ontario. *In* Jorgensen, Faulkner, and Mech (1970), pp. 32-38.

STARRETT, A.
1956. Pleistocene mammals of the Berends fauna of Oklahoma. Jour. Paleontol., 30:1187-1192.

STEBLER, A. M.
1944. The status of the wolf in Michigan. Jour. Mamm., 25:37-43.

STERNBERG, C. H.
1928. Extinct animals of California. Sci. Amer., 139:225-227.

STOCK, C.
1918. The Pleistocene fauna of Hawver Cave. Univ. California Publ. Bull. Dept. Geol., 10:461-515.
1929. A census of the Pleistocene mammals of Rancho La Brea, based on the collections of the Los Angeles Museum. Jour. Mamm., 10:281-289.
1938. A coyote-like wolf jaw from the Rancho La Brea Pleistocene. Bull. S. California Acad. Sci., 37:49-51.
1956. Rancho La Brea, a record of Pleistocene life in California. Los Angeles Co. Mus. Nat. Hist., Sci. Ser., no. 20, 83 pp.

STOCK, C., AND J. F. LANCE
1948. The relative length of limb elements in *Canis dirus*. Bull. S. California Acad. Sci., 47:79-84.

STOCK, C., J. F. LANCE, AND J. O. NIGRA
1946. A newly mounted skeleton of the extinct dire wolf from the Pleistocene of Rancho La Brea. Bull. S. California Acad. Sci., 45:108-110.

STOVALL, J. W., AND W. N. MCANULTY
1950. The vertebrate fauna and geologic age of Trinity River terraces in Henderson County, Texas. Amer. Midl. Nat., 44:211-250.

STRECKER, J. K.
1926. A check-list of the mammals of Texas. Baylor Univ. Bull., 29(3):1-48.

SURBER, T.
1932. The mammals of Minnesota. Minnesota Game and Fish Dept., St. Paul, 84 pp.

TAYLOR, R. W., C. I. COUNTS, III, AND S. MILLS
1976. Occurrence and distibution of the coyote, *Canis latrans*, Say, in West Virginia. Proc. West Virginia Acad. Sci., 48:3-4.

TEER, C. L.
1975. The eastern coyote. Maine Fish and Wildl., 17(4):9-12.

THENIUS, E.
1970. Einige jungpleistozane saugetiere (*Platygonus, Arctodus* und *Canis dirus*) aus dem Valsequillo, Mexico. Quartar, 21:57-66.

TROUESSART, E. L.
1911. Le loup de l'Inde (*Canis pallipes* Sykes), souche ancestrale du chien domestique. Compt. Rend. Acad. Sci. Paris, 152:909-913.

TROXELL, E. L.
1915. The vertebrate fossils of Rock Creek, Texas. Amer. Jour. Sci., ser. 4, 39:613-638.

ULMER, F. A.
1949. Recent records of coyotes in Pennsylvania and New Jersey. Jour. Mamm., 30:435-436.

VAN VALEN, L.
1964. Nature of the supernumary molars of *Otocyon*. Jour. Mamm., 45:284-286.

VANDERHOOF, V. L.
1933. Additions to the fauna of the Tehama upper Pliocene of northern California. Amer. Jour. Sci., ser. 5, 25:382-384.
1937. Critical observations on the Canidae in Cope's original collection from the Blanco of Texas. Proc. Geol. Soc. Amer., 1936, p. 389.

VANDERHOOF, V. L., AND J. T. GREGORY
1940. A review of the genus Aelurodon. Univ. California Publ. Bull. Dept. Geol., 25:143-164.

WARFEL, H. E.
1937. A coyote in Hampshire County, Massachusetts. Jour. Mamm., 18:241.

WEBB, S. D. (ed.)
1974a. Pleistocene mammals of Florida. University Presses of Florida, Gainesville, x+270 pp.

WEBB, S. D.
1974b. Chronology of Florida Pleistocene mammals. *In* Webb (1974a), pp. 5-31.

WEBB, W. S., AND R. S. BABY
1957. The Adena People. No. 2. Ohio Hist. Soc., Ohio State Univ. Press, xi+123 pp.

WEBSTER, D.
1964. Cherokee Cave bone deposit. Missouri Speleol., 16:79-86.

WEIGEL, R. D.
- 1962. Fossil vertebrates of Vero, Florida. Florida Geol. Surv. Spec. Publ., no. 10, vii+59 pp.

WEISE, T. F., W. L. ROBINSON, R. A. HOOK, AND L. D. MECH
- 1975. An experimental translocation of the eastern timber wolf. Audubon Conserv. Rept., no. 5, 28 pp.

WENDORF, F., A. D. KRIEGER, AND C. C. ALBRITTON
- 1955. The Midland discovery. Univ. Texas Press, viii+139 pp.

WETZEL, R. M., AND L. R. PENNER
- 1962. Coydog in Connecticut. Jour. Mamm., 43:109-110.

WHITACRE, D.
- 1948. The mysterious coyote pack of Ohio. Ohio Conserv. Bull., 12(3):29.

WHITNEY, J. D.
- 1879. The animal remains, not human, of the auriferous gravel series. Mem. Mus. Comp. Zool., 6:239-258.

WILLIAMS, C. T.
- 1962. Classification of the Borophaginae (Canidae). Unpublished M.A. thesis, Univ. Kansas, 103 pp.

WILLISTON, S. W.
- 1898. The Pleistocene of Kansas. Trans. Kansas Acad. Sci., 15:90-94.

WILSON, J.
- 1976. Bobcats, bears and coyotes . . . . are they here? Kentucky Happy Hunting Ground, 32(6):6-8.

WILSON, R. L.
- 1967. The Pleistocene vertebrates of Michigan. Papers Michigan Acad. Sci., Arts, and Letters, 52:197-234.

WILSON, R. W.
- 1933. Pleistocene mammalian fauna from the Carpenteria asphalt. Carnegie Inst. Washington Publ., 440:59-76.

WILSON, W. C.
- 1967. Food habits of the coyote, *Canis latrans*, in Louisiana. Unpublished M.S. thesis, Louisiana State Univ., ix+49 pp.

WING, E. S.
- 1963. Vertebrates from the Jungerman and Goodman sites near the east coast of Florida. Contrib. Florida State Mus., Soc. Sci., 10:51-60.

WOLFE, J. L.
- 1972. Wolves in Mississippi? Mississippi Game and Fish, 35(2):10-11.

WOLFE, M. L., AND D. L. ALLEN
- 1973. Continued studies of the status, socialization, and relationships of Isle Royale wolves, 1967 to 1970. Jour. Mamm., 54:611-635.

WOLFRAM, G.
- 1964. Coyotes: the silent invaders. Canadian Audubon, 26:112-115.

WOOD, N. A.
- 1922. The mammals of Washtenaw County, Michigan. Occas. Papers Mus. Zool. Univ. Michigan, no. 123, 23 pp.

WOOD, N. A., AND L. R. DICE
- 1924. Records of the distribution of Michigan mammals. Papers Michigan Acad. Sci., Arts, and Letters, 3:425-469.

WOODHOUSE, S. W.
- 1851. The North American jackal—*Canis frustror*. Proc. Acad. Nat. Sci., Philadelphia, ser. 1, 5:147-148.

WURSTER, D. H., AND K. BENIRSCHKE
- 1968. Comparative cytogenetic studies in the order *Carnivora*. Chromosoma, 24:336-382.

YOUNG, S. P.
- 1944. History, life habits, economic status, and control. Part I *in* Young, S. P., and E. A. Goldman, The wolves of North America, Amer. Wildl. Inst., Washington, D.C., pp. 1-385.
- 1946. The wolf in North American history. Caxton Printers, Ltd., Caldwell, Ohio, 149 pp.
- 1951. History, life habits, economic status, and control. Part I *in* Young, S. P., and H. H. T. Jackson, The clever coyote, Wildl. Mgmt. Inst., Washington, D.C., pp. 1-226.

Addendum

Subsequent to preparation of the galley proof of this paper, the following information came to my attention.

Canis lepophagus.—Bjork (1974) assigned a newly discovered specimen from the Wendell Fox pasture locality of the Rexroad fauna, Meade County, Kansas, to this species. Dalquest (1978) listed "*Canis* cf. *lepophagus*" from the Beck Ranch local fauna, Scurry County, Texas.

Canis latrans.—Parmalee, Munson, and Guilday (1978) reported specimens of "*C. latrans* Say—Coyote?" from the Harrodsburg Crevice, Monroe County, Indiana. Although a radiocarbon analysis made on bones from this site provided a date of 25,050±660 B.P., some of the faunal components suggested that a Sangamon age was more likely. Corner (1977) reported this species from a Rancholabrean fauna, 4.5 mi. W McCook, Red Willow County, Nebraska. Grayson (1977) reported "*Canis* cf. *latrans* Coyote" from zones dated 6,500-9,500 B.P. in the Dirty Shame Rockshelter, Malheur County, Oregon. Martin, Gilbert, and Adams (1977) listed this species from the late Pleistocene Natural Trap Cave, Big Horn County, Wyoming.

Canis rufus.—On 5 January 1978 the U.S. Fish and Wildlife Service made a second release of a pair of wild-caught red wolves on Bulls Island, Cape Romain National Wildlife Refuge, South Carolina. The animals apparently adapted well, and did not leave the vicinity of Bulls Island and a small adjacent island, until they were recaptured (unharmed) on 19 October and 1 November 1978 to terminate the experiment. In November 1978, the breeding colony at Tacoma, Washington and a second facility at Winnie, Texas contained a total of 31 wild-caught animals thought to be red wolves, 5 surviving young produced in the spring of 1977, and 15 young produced in the spring of 1978.

Canis lupus.—Corner (1977) reported this species from a Rancholabrean fauna, 4.5 mi. W McCook, Red Willow County, Nebraska. Martin, Gilbert, and Adams (1977) listed "*Canis* sp. (wolf)" from the late Pleistocene Natural Trap Cave, Big Horn County, Wyoming.

Canis familiaris.—Arredondo and Varona (1974) described *Cubacyon transversidens,* a new genus and species of canid from a Pleistocene site in western Cuba. Based on the published description and an examination of specimens of domestic dogs, E. Raymond Hall (Museum of Natural History, Univ. Kansas; pers. comm.) considers *Cubacyon transversidens* to be a synonym of *Canis familiaris,* and I agree with this assessment. Beebe (1978) reported a specimen of *C. familiaris* from the Old Crow River Basin of the northern Yukon, with a minimum age of 20,000 B.P., and observed: "The highly evolved morphology of the specimen suggests a much earlier time of domestication."

Canis dirus.—Parmalee, Munson, and Guilday (1978) reported specimens of "*Canis* cf. *dirus*" from the Harrodsburg Crevice, Monroe County, Indiana (see above paragraph on *C. latrans*), and from the Guy Wilson Cave, Sullivan County, Tennessee. According to Berta and Marshall (1978), fossils referable to *C. dirus* have been reported in South America from Talara, Peru; Tarija, Bolivia; and Muaco, Venezuela.

Literature Cited

Arredondo, O., and L. S. Varona
 1974. Nuevos genero y especie de mamifero (Carnivora: Canidae) del Cuaternario de Cuba. Poeyana (Havana), no. 131, 12 pp.

Beebe, B. F.
 1978. Two new Pleistocene mammal species from Beringia. Amer. Quat. Assoc., Abstr. 5th Bien. Mtg., p. 159.

Berta, A., and L. G. Marshall
 1978. South American Carnivora. *In* Westphal, F. (ed.), Fossilium catalogus, I: Animalia, W. Junk, The Hague, part 125, ix + 48 pp.

Bjork, P. R.
 1974. Additional carnivores from the Rexroad Formation (upper Pliocene) of southwestern Kansas. Trans. Kansas Acad. Sci., 76:24-38.

Corner, R. G.
 1977. A late Pleistocene-Holocene vertebrate fauna from Red Willow County, Nebraska. Trans. Nebraska Acad. Sci., 4:77-93.

DALQUEST, W. W.
1978. Early Blancan mammals of the Beck Ranch local fauna of Texas. Jour. Mamm., 59:269-298.

GRAYSON, D. K.
1977. Paleoclimatic implications of the Dirty Shame Rockshelter mammalian fauna. Tebiwa, no. 9, 26 pp.

MARTIN, L. D., B. M. GILBERT AND D. B. ADAMS
1977. A cheetah-like cat in the North American Pleistocene. Science, 195:981-982.

PARMALEE, P. W., P. J. MUNSON, AND J. E. GUILDAY
1978. The Pleistocene mammalian fauna of Harrodsburg Crevice, Monroe County, Indiana. Natl. Speleol. Soc. Bull., 40:64-75.

APPENDIX A

The following list provides details on samples used in multivariate analyses. Specimens are in the USNM unless otherwise indicated.

1. Canis lupus taken not later than 1925 in the mountainous region of western North America.

C. l. irremotus

ALBERTA.—25 mi. SE Lethbridge, 1.
IDAHO.—*Bannock Co.:* 10 mi. E Pocatello, 1; Tyhee Basin, 1. *Caribou Co.:* Soda Springs, 2. *Clark Co.:* Argora, 1. *Lemhi Co.:* Leadore, 1; 10 mi. S Leadore, 1.
MONTANA.—No precise locality, 1. *Beaverhead Co.:* Dillon, 1. *Carbon Co.:* Red Lodge, 1. *Carter Co.:* Ridge, 1. *Cascade Co.:* Belt, 1. *Powder River Co.:* Kruger, 1. *Rosebud Co.:* Ingomar, 1; Lame Deer, 1.
WYOMING.—*Campbell Co.:* Gilette, 1. *Converse Co.:* Glenrock, 1; Lost Springs, 1. *Fremont Co.:* Lenore, 1; Split Rock, 1. *Johnson Co.:* Barber, 1. *Sublette Co.:* Cora, 1; Pinedale, 1. *Sheridan Co.:* Arvada, 2. *Teton Co.:* Elk, 3; Kelly, 1. Yellowstone National Park, 1.

C. l. mogollonensis

ARIZONA.—No precise locality, 1. *Apache Co.:* Escudilla Mts., 3. *Greenlee Co.:* Clifton, 1; 15 mi. SW Alma, New Mexico, 1. *Maricopa Co.:* Aguila, 1. *Navajo Co.:* Cibecue, 1; Heber, 1.
NEW MEXICO.—*Catron Co.:* Datil Mts., 1; Gila National Forest, 6; Luna, 1; 15 mi. SE Reserve, 1. *Grant Co.:* head of Mimbres River, 1; Silver City, 1. *Sierra Co.:* Fairview, 1; Chlorida, 5; Monticello, 1. *Socorro Co.:* Magdalena, 1.

C. l. youngi

COLORADO.—*Mesa Co.:* no precise locality, 1; Glade Park, 2; West Creek, 1. *Pueblo Co.:* 25 mi. NW Pueblo, 1. *Rio Blanco Co.:* Piceance, 2; Sulphur, 1; Turman's Creek, 1.
NEW MEXICO.—*Rio Arriba Co.:* Abiquiu, 2; Canjilon, 1; Dulce, 2; El Vado, 1; Hayes, 2. *San Juan Co.:* La Plata, 1. *Sandoval Co.:* Cuba, 3; Senorita, 1. *Santa Fe Co.:* Lamy, 1. *Valencia Co.:* San Mateo, 1.
UTAH.—No precise locality, 1. *Box Elder Co.:* Grouse Creek, 1. *Duchesne Co.:* Duchesne, 1. *San Juan Co.:* 10 mi. NW Monticello, 2.
WYOMING.—No precise locality, 2. *Laramie Co.:* Federal, 2. *Sweetwater Co.:* Rock Springs, 1. Not located, Black Tail Creek, 1.

2. Canis latrans lestes taken not later than 1925 in the mountainous region of western North America.

COLORADO.—*Conejos Co.:* Bountiful, 1; Cenicro, 4; La Jara, 4; Rio Grande, 1. *Delta Co.:* Cedar Edge, 1; Grand Mesa, 2. *Garfield Co.:* Austin, 2; East Salt Creek, 2; Salt Creek, 1. *Grand Co.:* Kremmling, 4. *Larimer Co.:* Arkins, 3; Loveland, 1. *Mesa Co.:* Mesa, 1. *Moffatt Co.:* Craig, 2. *Park Co.:* South Park, 2; Tarryall, 3. *Rio Blanco Co.:* Piceance, 6. *Rio Grande Co.:* Monte Vista, 24. *Routt Co.:* Battle Creek, 4; Russell Springs, 4; Steamboat Springs, 2. *Summit Co.:* Gore Range, 1.
IDAHO.—*Ada Co.:* Boise, 1. *Bannock Co.:* no precise locality, 1; Chesterfield, 1; McCammon, 1; Pocatello, 3; Tyhee Basin, 2. *Bingham Co.:* Alridge, 2; Cerro Grande, 4; Ft. Hall, 2. *Blaine Co.:* Sawtooth National Forest, 3. *Bonneville Co.:* John Gray's Lake, 1. *Boundary Co.:* Schnoors, 1. *Canyon Co.:* Bowmont, 4. *Caribou Co.:* Preuss Mts., 2. *Cassia Co.:* Almo, 1; Oakley, 6. *Clark Co.:* Dubois, 1; Kilgore, 1; Medicine Lodge Creek, 4. *Custer Co.:* Bigfoot River, 2. *Elmore Co.:* Arrow Rock, 1. *Gooding Co.:* Gooding, 1. *Goodnow Co.:* Bliss, 1. *Idaho Co.:* Orangeville, 1; Rice Creek, 2; West Lake, 7; White Bird, 1. *Lemhi Co.:* Leadore, 2; Leesburg, 1; Salmon, 3. *Lewis Co.:* Forest, 1; Salmon River, 2. *Lincoln Co.:* Shoshone, 1. *Owyhee Co.:* Grand View, 1; Grassmere, 1; Hot Springs, 2; Three Creek, 8. *Payette Co.:* French, 1. *Pegram Co.:* Bear Lake, 3.

3. Canis familiaris.—50 (10 in KU, 6 in MCZ, 1 in ROM, 5 in UArk, 3 in USFWS).

4. Canis lupus from northern and western North America (other than as listed in 1 above).

C. l. alces

ALASKA.—Kachemak Bay, Kenai Peninsula, 2.

C. l. arctos

NORTHWEST TERRITORIES.—*Ellesmere Island:* Bear Peninsula, 1 (CNM); Eureka Sound, 6 (CNM); Griese Fjord, 2 (CNM); Hare Fjord, 1 (CNM); Slidre Fjord, Foshien Peninsula, 8 (6 in CNM). *Graham Island:* Norwegian Bay, 1 (CNM). *Prince Patrick Island:* Cherie Bay, 1; Mould Bay, 1 (CNM).

C. l. baileyi

ARIZONA.—*Cochise Co.:* Huachuca Mts., 1. *Pima Co.:* 5 mi. SE Arivaca, 1; Helvetia, 2.
NEW MEXICO.—*Dona Ana Co.:* Hatch, 1. *Grant Co.:* Cloverdale, 2; Hatchita, 4. *Hidalgo Co.:* Animas, 1; 30 mi. SE Animas, 2; 35 mi. SE Animas, 1; Animas Mts., 1; Animas Peak 1 (KU); San Luis Valley, 1.
TEXAS.—*Brewster Co.:* 10 mi. S Alpine, 1 (SR). *Jeff Davis Co.:* Fort Davis, 1. *Pecos Co.:* near Longfellow, 1 (SR).
CHIHUAHUA.—Colonia Garcia, 1; Colonia Juarez, 1; near corner adjoining Sonora, Arizona, and New Mexico, 3.
SONORA.—Sierra Pinto Mts., 1.

C. l. beothucus

NEWFOUNDLAND (Island).—No precise locality, 3 (2 in MCZ).

C. l. bernardi (including all specimens from Banks Island)

NORTHWEST TERRITORIES (all in CNM).—*Banks Island:* no precise locality, 1; North Adam River, 3; Big River, 1; Egg River, 1; 25 mi. E Sachs Harbor, 2.

C. l. crassodon

BRITISH COLUMBIA.—*Vancouver Island:* Quatsino Sound, 2.

C. l. fuscus

BRITISH COLUMBIA.—No precise locality, 1.
OREGON.—*Clackamas Co.:* Clackamas Lake, 1 (SD). *Curry Co.:* Rogue River, 2. *Douglas Co.:* Tiller, 2; Glide, 2. *Jackson Co.:* 25 mi. NE Ashland, 1; Peavine Mt., 3 (SD). *Lake Co.:* Sycan, 1. *Lane Co.:* 20 mi. S Oakridge, 1 (SD). *Linn Co.:* Cascadia, 2.
WASHINGTON.—*Jefferson Co.:* 22 mi. S Port Angeles, 1.

C. l. hudsonicus

NORTHWEST TERRITORIES.—Aberdeen Lake, 5 (CNM); Beaver Hill Lake, 3 (CNM); Cape Fullerton, 1 (AMNH); Hudson Bay, 1 (AMNH); Nueltin Lake, 1 (CNM); Red River, 1 (KU); head of Schultz Lake, 2; Simon's Lake, 1; Thelon River, 1; Wajer River, 1 (AMNH).

C. l. labradorius

NEWFOUNDLAND (LABRADOR).—Porcupine, 1.
QUEBEC (UNGAVA).—No precise locality, 1.

C. l. ligoni

ALASKA.—Conclusion Island, 1; Ketchikan, 1; Kuiu Island, 2; Kupreanof Island, 2; Prince of Wales Island, 2; Revillagigedo Island, 1; Wrangell, 6.

C. l. mackenzii

NORTHWEST TERRITORIES.—Amundsen Gulf, 1 (CNM); south side of Coronation Gulf, 1; Port Epworth Harbor, 1 (CNM); head of Hood River, 1 (CNM); Mackenzie Delta, 3 (CNM); Rae River, 1 (CNM).

C. l. manningi

NORTHWEST TERRITORIES.—*Baffin Island:* no pricise locality, 2 (1 in CNM, 1 in collection of Douglas H. Pimlott); Pangnirtung Fjord, 1 (CNM).

C. l. monstrabilis

NEW MEXICO.—No precise locality, 1. *Otero Co.:* Elk, 2; Sacramento Mts., 2; Mayhill, 1.
TEXAS.—*Crockett Co.:* Ozona, 2. *Culbertson Co.:* Guadalupe Mts., 1. *Jack Co.:* Fort Richardson, 1. *Kimble Co.:* 1. *Presidio Co.:* 40 mi. SW Marfa, 1. *Reagan Co.:* Big Lake, 1. *Upton Co.:* Rankin, 6. *Ward Co.:* Monahans, 1.

C. l. nubilus

MANITOBA.—Southeast of Carberry, 1 (CNM); Duck Mountain, 2; Riding Mountain National Park, 3 (CNM).
COLORADO.—*Bent Co.:* 3.
KANSAS.—*Gove Co.:* 3 mi. W Castle Rock, 1. *Trego Co.:* near Castle Rock, 1 (KU).
MINNESOTA.—*Becker Co.:* 25 mi. N Detroit Lakes, 1.
NEBRASKA.—Platte River, 3. *Kearny Co.:* Ft. Kearny, 3.
NEW MEXICO.—*Guadalupe Co.:* Santa Rosa, 1. *Lincoln Co.:* 40 mi. SE Corona, 1. *Socorro Co.:* Carthage, 3. *Torrence Co.:* Mountain Air, 1.
NORTH DAKOTA.—*Billings Co.:* Medora, 2. *Golden Valley Co.:* near Beach, 1.
OKLAHOMA.—Panhandle area, 1 (AMNH). *Comanche Co.:* Wichita Mountains National Wildlife Refuge, 1.
SOUTH DAKOTA.—No precise locality, 1. *Custer Co.:* Folsom, 1. *Harding Co.:* 20 mi. NE Buffalo, 1. *Meade Co.:* Faith, 1. *Pennington Co.:* Imlay, 1. *Ziebach Co.:* Red Elm, 1.
WYOMING.—*Converse Co.:* Douglas, 2. *Natrona Co.:* Natrona, 2.

C. l. occidentalis

ALBERTA.—Edmonton, 1; Simonette River, 1 (UAlb); 30 mi. N Whitecourt, 3 (UAlb); 50 mi. N Whitecourt, 3 (UAlb); Wood Buffalo National Park, 4 (AMNH).
BRITISH COLUMBIA.—Barking Horse River, 2 (KU); upper Henry River, 1.
NORTHWEST TERRITORIES.—Artillery Lake, 5 (4 in CNM); Aylmer Lake, 1 (AMNH); Fort Good Hope, 1; Fort Simpson, 1; Fort Smith, 1; Great Bear Lake, 1 (AMNH); 52 mi. up Keele River, 1 (CNM); Nahanni Butte, 10 (CNM); mouth of Netla River, 1 (CNM); Salt Plains, 5 (CNM); Slave River, 4 (CNM).
YUKON.—40 mi. SE Crow Base, 3; north fork McMillan River, 1; Pelly Lakes, 4; White River, 4 (ROM).

C. l. orion

GREENLAND.—No precise locality, 2 (AMNH).

C. l. pambasileus

ALASKA.—No precise locality, 1. Anaktuvak Pass, 7 mi. N Tolugak, 1; Big Delta River, 1; Cold Bay, 1; Fairbanks, 1; 100 mi. N Fairbanks, 1; Farewell Mts., 1; Gold Creek (near head, above Curry), 1; Jarvis Creek, 1; upper John River, 7; Little Delta River, 1; Mt. Hayes, 3; Nome, 1; Savage River, 1; Sushana River, 2; Tanana River, 2; Teklanika River, 2; Teller, 1; Tolugak Lake, 1; Yukon River, 35 mi. below Beaver, 1.
YUKON.—No precise locality, 1; Hoole Canyon, 1.

C. l. tundrarum

ALASKA.—No precise locality, 1; Noatak River, 2; Pitmega River, Cape Sabine, 1; Point Barrow, 1; Umiat, 2; upper Meade River, 1 (UCMVZ); Wahoo Lake, Brooks Range, 1 (KU).

C. l. youngi

CALIFORNIA.—*San Bernardino Co.:* 12 mi. W Lanfair, 1 (UCMVZ).

5. Canis latrans from northern and western North America (other than as listed in part 2).

C. l. incolatus
ALASKA.—Big Delta River, 12; Copper River Flats, 1; Eagle River, 4; Fairbanks, 2; Mt. Hayes, 4; Tanana, 1.

C. l. latrans
WYOMING.—*Albany Co.:* Jelm, 1; Laramie, 22; Red Mts., 1. *Carbon Co.:* Shirley, 1. *Converse Co.:* Douglas, 4. *Crook Co.:* Manville, 4; Sundance, 4. *Laramie Co.:* Federal, 4. *Natrona Co.:* Casper, 1. *Sheridan Co.:* Arvada, 1.

C. l. mearnsi
ARIZONA.—*Apache Co.:* Marsh Lake, 1; Springerville, 1. *Coconino Co.:* Anderson Mesa, 1; Bright Angel Spring, 1; Flagstaff, 1; Fredonia, 3; Kaibab National Forest, 1; Ryan, 6; Tuba, 4. *Graham Co.:* Chiricahua Ranch, 1. *Mojave Co.:* Trumbull Mts., 1. *Navajo Co.:* Antelope Springs, 1; Ft. Apache, 4. *Yuma Co.:* Gila Mts., 1; Tinajas Altas, 1; Tule Tanks, 1.

C. l. texensis
NEW MEXICO.—*Bernalillo Co.:* Isleta, 1. *Eddy Co.:* Salt Valley, 2. *Lincoln Co.:* Callo Canyon, 3. *Otero Co.:* Cienega, 3; Lincoln National Forest, 3. *San Juan Co.:* Fruitland, 1. *San Miguel Co.:* Pecos, 1. *Santa Fe Co.:* Lamy, 2. *Socorro Co.:* Carthage, 6; San Andres Mts., 1. *Torrance Co.:* Manzano Mts., 1; Mesa Jiminez, 1.

6. Suspected hybrids.

Canis lupus x Canis latrans
ARIZONA.—Not located, Lanks, 1.
CHIHUAHUA.—Colonia Garcia, 1 (MCZ).
VERACRUZ.—Orizaba, 1 (MCZ).
ONTARIO.—Captives, 2 (ROM). *Lanark Co.:* Sherbrooke, 1 (ROM). *Nipissing District:* Preston, 1 (ROM).
QUEBEC.—*Gatineau Co.:* northern part, 1 (QWS); central part, 1 (QWS); Gracefield, 1 (CNM, originally identified as *C. latrans thamnos*). *Papineau Co.:* Montebello, 1 (ROM). *Pontiac Co.:* Head Lake, 1 (CNM).

Canis lupus x Canis familiaris
MICHIGAN.—*Luce Co.:* McMillan, 1 (UMMZ). *Schoolcraft Co.:* Cusino, 1.
NEW MEXICO.—*Otero Co.:* Sacramento Mts., 2.

7. Canis lupus lycaon.

Western Group
ONTARIO.—*Algoma District:* Batchwana Bay, 1 (UCMVZ); McMahon Twp., 1 (ROM). *Cochrane Dist.:* Kapukasing, 1 (collection of Douglas H. Pimlott). *Kenora Dist.:* no precise locality, 1 (ROM); Ball Lake, 1 (CNM); Eagle Lake, 1 (ROM); 100 mi. W Fort William, 1 (CNM); Kenora, 1 (ROM); Whitefish Bay, 2 (ROM). *Parry Sound Dist.:* Burton Twp., 1 (ROM); Carling Twp., 1 (ROM). *Rainy River Dist.:* Quetico, 3 (UI). *Thunder Bay Dist.:* Hurkett, 1 (ROM); Killala Lake, 2 (ROM); Lake Leopard, 1 (ROM); Lape Nipigon, 2 (CNM); north shore of Lake Superior, 1 (CNM); Silver Islet, 1 (ROM).
MICHIGAN.—*Alger Co.:* southern part, 1 (UMMZ); Grand Marais, 1 (UMMZ); 14 mi. SW Grand Marais, 1 (UMMZ); 25 mi. NE Munising, 1 (UMMZ). *Baraga Co.:* Pheshika River, 1 (UMMZ); Sec. 6, T50N, R31W, 1 (MSU). *Chippewa Co.:* north shore of Whitefish Bay, 1 (UMMZ); Sec. 35, T47N, R5W, 2 (MSU). *Delta Co.:* West Escanaba River, 1. *Dickinson Co.:* no precise locality, 1; Randville, 1; West Escanaba River, 1. *Gogebic Co.:* Iron River, 1 (UMMZ); Marinesco, 1 (UMMZ); Presque Island, 2(UMMZ); 7 mi. N Watersmeet, 1 (MSU). *Houghton Co.:* Kenton, 1. *Luce Co.:* north of Newberry, 1. *Marquette Co.:* 30 mi. NW Marquette, 2. *Ontonagon Co.:* Calderwood, 1. *Schoolcraft Co.:* 1.
MINNESOTA.—*Beltrami Co.:* Red Lake National Wildlife Refuge, 1 (UMinn). *Cook Co.:* no precise locality, 3 (AMNH); near Dunn Lake, 2 (UMinn); Horland, 1 (UMinn). *Koochiching Co.:* 2 (1 in TM, 1 in UMinn). *Lake Co.:* Clearwater Lake, 1; Eskwagama Lake, 1; Hart Lake, 1; Horse River, 1; South Fowl Lake, 1. *Lake of the Woods Co.:* Baudette, 1 (UMinn); 12 mi. S Williams, 1 (UMinn). *St. Louis Co.:* Duluth, 1 (AMNH); Ely, 6 (UMinn); Four Town Lake, 1 (UI). *Sherburne Co.:* Elk River, 1.
WISCONSIN.—*Vilas Co.:* Eagle River, 1.

Eastern Group
ONTARIO.—*Nipissing Dist.:* Algonquin Provincial Park, 7 (3 in CNM, 3 in ROM); Bishop Twp., 1 (ROM); Clancy Twp., 1 (KU); Lake Nipissing, 1 (ROM); Preston, 1 (ROM); Whitney, 1 (UCMVZ). *Peterborough Co.:* north of Apsley, 1 (CNM). *Renfrew Co.:* Dacre, 1 (ROM).
QUEBEC.—Southern part, no precise locality, 2 (QWS). *Gatineau Co.:* Aylwin, 1 (CNM); Lucerne, 2 (UCMVZ). *Labelle Co.:* Boyer, 1 (QWS); Lacoste, 1 (QWS); Mont Laurier, 1 (QWS); Nominingu, 1 (QWS); Ste. Veronique, 1 (QWS); Val-Barrette, 1 (QWS). *Papineau Co.:* Montebello, 2 (ROM). *Pontiac Co.:* near Cabonga Reservoir, 2 (QWS); Jim's Lake, 1 (CNM). *Temiscamingue Co.:* 40 mi. NE Mattawa, 1.

8. Canis latrans thamnos.
MANITOBA.—Carman, 4; Duck Mountain, 1.
ONTARIO.—*Algoma Dist.:* Dean Lake, 1 (ROM); Prince, 1 (ROM); Tarbutt, 4 (ROM); Wolford, 1 (ROM). *Greg Co.:* Markdale, 1 (ROM). *Huron Dist.:* Zurich, 1 (ROM). *Kenora Dist.:* Oxdrift, 1 (ROM). *Kent Co.:* Chatham, 1 (ROM). *Lambton Co.:* Thedford, 1 (CNM). *Lanark Co.:* Sherbrooke, 1 (ROM). *Nipissing Dist.:* Algonquin Provincial Park, 2 (CNM). *Norfolk Co.:* 1 (ROM). *Parry Sound Dist.:* Monteith, 1 (ROM). *Peterborough Co.:* Lakefield, 1 (CNM). *Rainy River Dist.:* Pinewood, 1 (ROM).
QUEBEC.—No precise locality, 2 (QWS). *Beauce Co.:* Beauceville, 1 (QWS). *Charlevoix*

Co.: Baie St. Paul, 1 (QWS). *L'Islet Co.:* St. Aubert, 1 (QWS). *Maskinonge Co.:* St. Leon, 1 (QWS). *Portneuf Co.:* Valcartier, 1 (QWS).

ILLINOIS.—*Lake Co.:* Camp Logan, 1 (FM). *McClean Co.:* LeRoy, 1. *Marshall Co.:* 9 mi. W Henry, 1.

INDIANA.—*Clinton Co.:* Jefferson, 1 (PUWL). *Jasper Co.:* McCoysburg, 1. *Newton Co.:* 5 mi. S Roselawn, 1 (PUWL). *Tippecanoe Co.:* West Point, 1 (PUWL).

IOWA.—*Adair Co.:* Richland, 2 (KU). *Appanoose Co.:* Moravia, 1 (KU). *Monroe Co.:* 2 mi. N Avery, 2 (KU).

MICHIGAN.—*Alcona Co.:* Aldair, 1 (UMMZ). *Alger Co.:* Miners River, 1. *Baraga Co.:* no precise locality, 2 (MSU); Baraga, 1 (UMMZ). *Barry Co.:* 1 (UMMZ). *Cheboygan Co.:* Beaugrand, 1 (UMMZ). *Chippewa Co.:* Brimley, 1 (UMMZ); 7 mi. NW Pickford, 1 (UMMZ); Race, 1 (UMMZ). *Clinton Co.:* St. Johns, 1 (MSU). *Crawford Co.:* Hanson Game Refuge, 1 (UMMZ). *Delta Co.:* Bark River, 1; Rapid River, 1. *Dickinson Co.:* Cedar River, 1 (UMMZ). *Gogebic Co.:* Ironwood, 2 (UMMZ); Montreal River, 1 (UMMZ). *Houghton Co.:* Isle Royale, 3 (UMMZ). *Ingham Co.:* 1 (MSU). *Iron Co.:* 3 (MSU). *Jackson Co.:* Liberty, 1 (UMMZ). *Marquette Co.:* Negaunee, 2; Yalmar, 1 (UMMZ). *Menominee Co.:* Cedar River, 2 (UMMZ); Dagett, 1; Ingalls, 4 (UMMZ); Michigamee River, 1; Whitney, 2 (UMMZ); Wilson, 2 (UMMZ). *Montcalm Co.:* 1 (MSU). *Ontonagon Co.:* no precise locality, 1 (UMMZ); Ewen, 1 (UMMZ). *St. Clair Co.:* 1 (MSU). *Schoolcraft Co.:* no precise locality, 2 (MSU); Manistique, 1 (UMMZ). *Washtenaw Co.:* Dexter, 1 (UMMZ). Not located, Warheim, 1 (UMMZ).

MINNESOTA.—*Beltrami Co.:* no precise locality, 4 (UMinn); Red Lake National Wildlife Refuge, 1 (UMinn). *Isanti Co.:* 1 (UMinn). *Lake Co.:* Fernberg, 1 (UMinn). *Lake of the Woods Co.:* no precise locality, 5 (UMinn); Norris Camp, 2 (UMinn). *Pennington Co.:* 1 (UMinn). *Pine Co.:* 1 (UMinn). *Sherburne Co.:* Elk River, 5.

NORTH DAKOTA.—*Benson Co.:* Ft. Totten, 1; Sully Hill National Park, 1.

WISCONSIN.—*Ashland Co.:* Basswood Island, Apostle Islands, 1. *Forest Co.:* Crandon, 4 (MSU); Wabeno, 1 (MSU). *Iron Co.:* Kenosa, 1 (FM). *Vilas Co.:* Eagle River, 1. *Walworth Co.:* Delavan, 1.

9. **Wild Canis from the Northeastern United States.**

MAINE.—*Franklin Co.:* Rangley, 1. *Kennebec Co.:* Monmouth, 1.

MASSACHUSETTS.—*Berkshire Co.:* Otis, 1 (MCZ). *Franklin Co.:* Colrain, 1 (MCZ); Leyden, 1 (MCZ).

NEW HAMPSHIRE.—*Coos Co.:* Lancaster, 1 (MCZ); Stewartstown, 1 (MCZ). *Hillsborough Co.:* Temple, 2 (MCZ). *Merrimack Co.:* Boscawen, 1 (MCZ). *Sullivan Co.:* Croydon, 1 (MCZ).

NEW YORK.—No precise locality, 1 (NYEC). *Franklin Co.:* Faust, 1 (NYEC); Santa Clara, 2 (NYEC). *Lewis Co.:* 4 (NYEC). *Oneida Co.:* Hawkinsville, 2 (NYEC); Woodgate, 1 (NYEC). *Oswego Co.:* Fulton, 1 (NYEC). *Schenectady Co.:* 1 (NYEC). *Yates Co.:* 1.

PENNSYLVANIA.—*Clearfield Co.:* Clearfield, 1. *Potter Co.:* 1.

VERMONT.—No precise locality, 1. *Addison Co.:* Granville, 1 (VFG). *Chittenden Co.:* Shelburne, 1 (VFG). *Orange Co.:* Brookfield, 1 (VFG). *Orleans Co.:* Barton, 2 (1 in VFG); Glover, 1 (VFG); Jay, 1 (VFG); Troy, 2 (VFG). *Rutland Co.:* no precise locality, 1 (VFG); Middletown Springs, 1 (VFG); Sudbury, 1 (VFG). *Washington Co.:* Berlin, 1 (VFG); Montpelier, 1 (VFG). *Windham Co.:* Brookline, 2 (MCZ); Wardsboro, 1 (MCZ).

10. **Canis rufus gregoryi**, 1919-1929 (for data on earlier material see table 2).

ARKANSAS.—*Boone Co.:* Bergman, 1. *Cleburne Co.:* Almond, 1. *Dallas Co.:* Carthage, 1. *Garland Co.:* Crystal Springs, 1; Lonsdale, 2 (1 in MCZ). *Marion Co.:* Mull, 1. *Newton Co.:* Fallsville, 7; Lurton, 2. *Perry Co.:* Ava, 1; Cedar, 1. *Polk Co.:* Egger, 1; 12 mi. NE Egger, 2; 10 mi. W Egger, 1; Mena, 1; Shady, 1. *Pope Co.:* Mill Creek, 2; Simpson, 5; Solo, 1. *Pulaski Co.:* Fernsdale, 5; Pinnacle, 2. *Saline Co.:* Isaac, 6. *Scott Co.:* Blue Ball, 4; Cardiff, 2; 4 mi. S Parks, 1. *Yell Co.:* 8 mi. NW Aly, 1; Onyx, 8; Stillwater, 3.

LOUISIANA.—*Beauregard Parish:* near Sabine River, 2. *Madison Pa.:* 1.

MISSOURI.—*Carter Co.:* Barren, 4. *Crawford Co.:* Cook Station, 3. *Howell Co.:* West Plains, 2. *Iron Co.:* Arcadia, 2. *Ripley Co.:* Gatewood, 3. *Stone Co.:* 3. *Texas Co.:* Tyrone, 1. *Wayne Co.:* Upalika, 1.

OKLAHOMA.—*Le Flore Co.:* Octavia, 1; Page, 3; Talihina, 1. *McCurtain Co.:* Bethel, 3; Broken Bow, 7; Sherwood, 4; Smithville, 7. *Pushmataha Co.:* Cedar Creek, 1; Fewell, 1; Nashoba, 1.

11. **Southeastern specimens** that suggest hybridization with *Canis familiaris*.

ARKANSAS.—*Pope Co.:* 4 mi. S Raspberry, 1.

LOUISIANA.—Northern part, no precise locality, 1 (LPI). *Jackson Pa.:* 1 (LPI). *Winn Pa.:* Sikes, 1.

MISSOURI.—*Iron Co.:* 4 mi. S Sabula, 1.

TEXAS.—*Lavaca Co.:* 20 mi. S Hallettsville, 1. *Van Zandt Co.:* 1.

12. **Specimens**, originally identified as *C. rufus gregoryi*, with short greatest lengths.

ARKANSAS.—*Marion Co.:* Mull, 1. *Newton Co.:* Fallsville, 2. *Perry Co.:* Ava, 1; 8 mi. W Wye, 2. *Pope Co.:* Simpson, 1. *Pulaski Co.:* Fernsdale, 1. *St. Francis Co.:* Forrest, 1.

OKLAHOMA.—*Le Flore Co.:* Octavia, 1; Page, 1. *McCurtain Co.:* Bethel, 1; Broken Bow, 1; Smithville, 2.

13. *Canis latrans*, pre-1930, southern Missouri.—*Carter Co.:* Barren, 2. *Phelps Co.:* Rolla, 5. *Saline Co.:* 4 mi. N Napton, 1 (MCZ). *Texas Co.:* Tyrone, 1.

14. Pre-1930 specimens originally identified as *Canis rufus rufus*.

ARKANSAS.—*Newton Co.:* Boxley, 1.
OKLAHOMA.—*Atoka Co.:* near Atoka, 2. *Garvin Co.:* Cherokee Town, 40 mi. N Ardmore, 1. *Tulsa Co.:* Red Fork, 2.
MISSOURI.—*Stone Co.:* Reeds Springs, 2.

15. Pre-1930 specimens originally identified as *Canis latrans*.

OKLAHOMA.—*Canadian Co.:* Calumet, 5. *Comanche Co.:* Cache, 3. *Creek Co.:* Manford, 1. *Custer Co.:* Anthon, 1; Butler, 5. *Tillman Co.:* Frederick, 9. *Tulsa Co.:* Red Fork, 1.
TEXAS.—*Hemphill Co.:* 2.

16. *Canis rufus rufus*, 1900 and 1904, coastal Texas.—*Calhoun Co.:* O'Connorsport, 4; 7 mi. SW Port Lavaca, 2. *Colorado Co.:* Frelsburg, 1. *Liberty Co.:* 6 mi. N Dayton, 1.

17. *Canis latrans texensis*, pre-1930, southern Texas.—*Frio Co.:* Frio Town, 1; 11 mi. W Frio Town, 1; 20 mi. W Frio Town, 2; 8 mi. SW Frio Town, 4; 9 mi. S Moore, 1; Pearsall, 7; 5 mi. E Pearsall, 1; 20 mi. W Pearsall, 3. *Nueces Co.:* Corpus Christi, 27; 45 mi. SW Corpus Christi, 3; Nueces Bay, 2; San Diego, 1. *Uvalde Co.:* Sabinal, 1; 10 mi. N Sabinal, 1; 5 mi. S Sabinal, 1. *Zavala Co.:* 12 mi. NE Batesville, 1.

18. *Canis latrans texensis*, pre-1930, western Texas.—*Brewster Co.:* Alpine, 3. *Coke Co.:* 10 mi. N Water Valley, 1. *Crockett Co.:* Ozona, 3; 9 mi. W Ozona, 1; 12 mi. NW Ozona, 5. *Pecos Co.:* Sheffield, 1. *Reagan Co.:* Big Lake, 4; 3 mi. N Big Lake, 1; 25 mi. E Big Lake, 2; 12 mi. S Big Lake, 2. *Sterling Co.:* Broome, 1; Sterling City, 3; 30 mi. S Sterling City, 1. *Upton Co.:* Rankin, 7; 10 mi. SW Rankin, 9. *Nolan Co.:* Sweetwater, 1.

19. *Canis latrans texensis*, pre-1930, Tom Green County, Texas.—Carlsbad, 1; 6 mi. NE Carlsbad, 1; 15 mi. NE Carlsbad, 2; Christoval, 4; 6 mi. NE Christoval, 1; 10 mi. NE Christoval, 1; 15 mi. NE Christoval, 2; 20 mi. NE Christoval, 1; Mereta, 1; San Angelo, 25; 15 mi. W San Angelo, 2; Water Valley, 9.

20. Specimens from central Texas, pre-1930.—*Blanco Co.:* Blanco, 2; Round Mt., 1. *Burnet Co.:* Burnet, 1; 5 mi. E Fairland, 1; Marble Falls, 6; 6 mi. S Marble Falls, 1. *Coleman Co.:* 16 mi. N Coleman, 1. *Concho Co.:* 5 mi. N Pasche, 1. *Edwards Co.:* Nueces River, 1. *Gillespie Co.:* 2. *Kerr Co.:* no precise locality, 4; Kerrville, 1. *Llano Co.:* no precise locality, 4; Baby Head, 1; 22 mi. S Bird Range, 1; Castell, 7; Click, 2; Llano, 7; 20 mi. N Llano, 1; 15 mi. E Llano, 1; 20 mi. S Llano, 3; 7 mi. NW Llano, 2; Valley Springs, 2. *McCulloch Co.:* Brady, 3; 13 mi. SW Brady, 1; 5 mi. SE Doole, 1. *Menard Co.:* Callan, 1; Ft. McKavett, 1; Menard, 10. *San Saba Co.:* Cherokee, 2. *Sutton Co.:* Sonora, 1; 25 mi. W Sonora, 2.

21. *Canis rufus*, 1930's-1950's.

C. r. gregoryi

ALABAMA.—*Sumter Co.:* Livingston, 1.
ARKANSAS.—Union-Columbia county line, 1.
LOUISIANA.—*La Salle Pa.:* Little River, 1 (LSUMZ). *Madison Pa.:* Tallulah Reservation, 2 (LUSMZ). *Terrebonne Pa.:* near Houma, 1 (LSUMZ). *Winn Pa.:* 3.
MISSISSIPPI.—*Harrison Co.:* Biloxi, 1 (AMNH).
OKLAHOMA.—*McCurtain Co.:* near Battiest, 2 (UArk).
TEXAS.—*Hardin Co.:* no precise locality, 1; Honey Island, 1 (UAriz); Kountze, 1. *Newton Co.:* 1. *Polk Co.:* southern part, 1; Carmona, 1 (UCMVZ); near Wakefield, 2.

C. r. rufus

TEXAS.—*Brazoria Co.:* 12 mi. S, 4 mi. E Alvin, 1 (KU); Angleton, 1; 9 mi. NE Angleton, 2; 5 mi. E Angleton, 1; 12 mi. E Angleton, 1. *Brazos Co.:* 15 mi. S Bryan, 1. *Harris Co.:* Genoa, 1. *Liberty Co.:* Cleveland, 1; 1.5 mi. N Rye, 2. *Madison Co.:* 11 mi. SE Madisonville, 2. *Montgomery Co.:* Porter, 2; Security, 2. *Walker Co.:* New Waverly, 1.

22. Specimens from the central coast of Texas, 1936-1942.—*Aransas Co.:* Aransas National Wildlife Refuge, 5. *Refugio Co.:* 22 mi. E Refugio, 1; 12 mi. S Tivoli, 1; 7 mi. S Woodsboro, 1. *Victoria Co.:* Bloomington, 1; 6 mi. S Bloomington, 1.

23. Specimens from northern Texas, 1930-1942.—*Eastland Co.:* Cisco, 1. *Jack Co.:* Henry Lewis Ranch, 2; 25 mi. NW Jacksboro, 1. *Palo Pinto Co.:* no precise locality, 1; 6 mi. NE Graford, 1. *Parker Co.:* 1. *Shackelford Co.:* 2. *Throckmorton Co.:* 5. *Wilbarger Co.:* 16 mi. SE Vernon, 1. *Young Co.:* 6 mi. NE Murray, 3 (also examined, 3 specimens from near San Antonio, Bexar Co.).

24. *Canis latrans frustror*, Wichita Mountains National Wildlife Refuge, Comanche County, southwestern Oklahoma, 1933-1942.—47.

25. Specimens from central and northeastern Oklahoma, 1932.—*Cherokee Co.:* 1. *Cleveland Co.:* Noble, 3. *Osage Co.:* 2 (also examined, one specimen of *C. r. rufus* from Redden, Atoka Co.).

26. Specimens from southern Missouri, 1941-1942. —*Christian Co.:* 1. *Crawford Co.:* 1. *Taney Co.:* 3. *Texas Co.:* 1. *Vernon Co.:* 9 (also examined, one specimen from Dade Co., collected 1932; one from 3 mi. N Thomasville, Oregon Co. (UCMVZ), collected 1942; and one from 5 mi. N Gainesville, Ozark Co. (UCMVZ), collected 1941).

27. Specimens from Arkansas, 1930-1951.—No precise locality, 1 (UArk). *Benton Co.:* Cherokee

City, 3; Siloam Springs, 1; Springtown, 1. *Howard Co.*: Umpire, 1. *Lawrence Co.*: 1. *Stone Co.*: State Game Refuge, 1 (UArk). *Washington Co.*: Summers, 2; Devil's Den State Park, 2 (UArk).

28. Specimens from southeastern Oklahoma, post-1960.—*Bryan Co.*: 4. *Choctaw Co.*: 8. *McCurtain Co.*: 6. *Pushmataha Co.*: 7.

29. Specimens from northern Arkansas, post-1960 (all in UArk).—*Conway Co.*: 8. *Franklin Co.*: 2. *Newton Co.*: 2. *Pope Co.*: 1. *Van Buren Co.*: 7.

30. Specimens from southern Arkansas, post-1960. —*Calhoun Co.*: 1 (UArk). *Chicot Co.*: 3. *Clark Co.*: 1. *Hempstead Co.*: 26 (15 in UArk). *Hot Springs Co.*: 4 (UArk). *Howard Co.*: 1 (UArk). *Little River Co.*: 14. *Miller Co.*: 6. *Nevada Co.*: 4 (UArk). *Sevier Co.*: 3 (UArk).

31. Specimens from Louisiana, post-1960.—*Beauregard Pa.*: near Merryville, 8. *Bienville Pa.*: 2 (1 in LPI, 1 in LSUMZ). *Bossier Pa.*: 1 (LPI). *Concordia Pa.*: Ferriday, 1. *De Soto Pa.*: 3 (LPI). *East Carroll Pa.*: 4 mi. N Transylvania, 1 (LPI). *Jackson Pa.*: 7 (LPI). *Natchitoches Pa.*: 4 (LPI). *Red River Pa.*: 1. *St. Landry Pa.*: Thistlewaite Game Management Area, 3 (LSUMZ). *Union Pa.*: 3 mi. S Farmerville, 1 (LPI); 2 mi. N Farmerville, 1 (LPI). *Webster Pa.*: 1 (LPI). *West Baton Rouge Pa.*: 2 mi. W Addis, 1 (LSUMZ). *Winn Pa.*: 2 mi. S Brewster's Mill, 5 (LPI); 4 mi. E Dodson, 1 (LPI) (also examined, one specimen, apparently *C. rufus gregoryi*, from near Washington, St. Landry Pa., collection of Douglas H. Pimlott).

32. Specimens from inland east Texas, post-1960. —*Bell Co.*: 2. *Bosque Co.*: 2. *Bowie Co.*: 3. *Cherokee Co.*: 12. *Collin Co.*: 2. *Delta Co.*: 5. *Denton Co.*: 16. *Freestone Co.*: 7. *Grayson Co.*: 14. *Hamilton Co.*: 3. *Hopkins Co.*: 4. *Hunt Co.*: 11. *Johnson Co.*: 8. *Lamar Co.*: 7. *Leon Co.*: 10. *Limestone Co.*: 2. *Milam Co.*: 5. *Morris Co.*: 3. *Rusk Co.*: 2. *Smith Co.*: 4.

33. Specimens from the central coast of Texas, post-1960.—*Austin Co.*: 7 mi. NW Sealy, 2 (MSU). *Calhoun Co.*: 7. *Colorado Co.*: 14 (MSU). *Fort Bend Co.*: 3. *Lavaca Co.*: 7. *Matagorda Co.*: 8. *Victoria Co.*: 1 (USFWS) (also examined, one specimen, apparently *C. rufus rufus*, from near Armstrong, Kenedy Co., collection of Russell E. Mumford).

34. Specimens from the vicinity of the Addicks Reservoir, Harris Co., Texas, post-1960.—12 (7 in USFWS).

35. Specimens from the vicinity of the Clemens Prison Farm, western Brazoria Co., Texas, post-1960. —31 (11 in USFWS).

36. Specimens from the eastern part of Brazoria Co., Texas, post-1960.—15 mi. S Alvin, 1 (USFWS); 5 mi. E Angleton, 5; 7 mi. N Angleton, 1 (USFWS); Graham Ranch, 1; near Hoskins Mound, 10 (7 in USFWS); Liverpool, 1; Stringfellow Ranch, 1.

37. Specimens from the vicinity of the Big Thicket southeastern Texas, post-1960.—*Chambers Co.*: 7 mi. E Baytown, 4 (USFWS). *Jasper Co.*: near New Blox, 1 (USFWS). *Liberty Co.*: no precise locality, 3; 1 mi. S Ames, 1 (USFWS); 4 mi. S Ames, 1 (USFWS); 5 mi. S Dayton, 2 (USFWS); 2 mi. E Devers, 2 (USFWS); 5 mi. N Liberty, 2 (USFWS); 15 mi. E Liberty, 2 (USFWS); 3 mi. S Raywood, 1 (USFWS); 5 mi. S Raywood, 3 (USFWS). *Tyler Co.*: near Fred, 2 (USFWS).

38. *Canis rufus gregoryi*, southeastern Texas, 1963-1970.—*Chambers Co.*: Anahuac National Wildlife Refuge and vicinity, 6 (2 in USFWS); Barrows Ranch, 1 (UO); Canada Ranch, 2 (USFWS); Double Bayou, 4; Logan Ranch, 2; Monroe City, 1; Smith Point, 1 (USFWS). *Jefferson Co.*: near Port Arthur, 2.

APPENDIX B

This appendix provides measurements for some of the key series used in multivariate analysis, and for some of the fossil specimens examined. The numbered parts of the appendix (left margin) are the same as referred to in the text. The numbers along the tops of the columns correspond to the numbers of the 15 measurements described below. If no sample size (n) is indicated, or if an asterisk (*) follows the sample size, then the figures shown are actual measurements of individuals. Otherwise, the five horizontal rows under the designation and sample size (n) of the series are mean, lower extreme, upper extreme, standard deviation, and coefficient of variation. Figures in parentheses, following the sample size of most series for which sex is designated, represent the number of specimens in the sample that were unknown as to sex, but which were judged to belong to the category indicated.

Descriptions of Measurements

1. *Greatest length.*—Length from anterior tip of premaxillae to posterior point of inion.
2. *Zygomatic width.*—Greatest distance across zygomata.
3. *Braincase width.*—Maximum breadth of braincase across level of parietotemporal sutures.
4. *Alveolar length of maxillary toothrow.*—Distance from anterior edge of alveolus of P1 to posterior edge of alveolus of M2.
5. *Maximum crown width across upper cheek teeth.*—Greatest breadth between outer sides of most widely separated upper teeth (P4 or M1).
6. *Palatal width at P1.*—Minimum width between inner margins of alveoli of first upper premolars.
7. *Width at C1.*—Greatest breadth across maxillae at outer edges of alveoli of canines.
8. *Width of frontal shield.*—Maximum breadth across postorbital processes of frontals.
9. *Postorbital constriction.*—Least width across frontals at constriction behind postorbital processes.
10. *Length from toothrow to bulla.*—Minimum distance from posterior edge of alveolus of M2 to depression in front of bulla at base of muscular process.
11. *Height from maxillary toothrow to orbit.*—Minimum distance from outer alveolar margin of M1 to most ventral point of orbit.
12. *Depth of jugal.*—Minimum depth of jugal anterior to postorbital process, at right angle to its anteroposterior axis.
13. *Diameter of C1.*—Maximum anteroposterior width of upper canine at base of enamel.
14. *Crown length of P4.*—Maximum anteroposterior length of crown measured on outer side.
15. *Crown width of M2.*—Maximum transverse diameter from outermost point to innermost point of crown.

MEASUREMENTS

	1	2	3	4	5	6	7	8	9	10	11	12	13	14	15
1. *C. familiaris*, n=50															
	217.2	112.4	58.95	70.22	68.12	29.76	41.64	60.94	39.21	60.24	33.61	15.25	11.21	19.28	10.91
	151.0	84.0	50.5	52.5	51.5	21.5	30.0	40.5	32.2	33.6	20.5	10.1	8.4	14.4	7.7
	285.0	154.0	65.0	88.0	85.5	42.3	59.0	87.4	44.8	88.0	53.5	23.6	14.0	22.7	13.0
	30.88	12.91	3.07	8.19	7.33	4.96	6.42	9.78	3.17	11.24	6.78	2.67	1.47	1.66	1.27
	14.27	11.48	5.21	11.66	10.76	16.67	15.42	16.09	8.08	18.66	20.17	17.51	13.11	8.61	11.64
2. Total sample of northern and western *C. lupus*, male, n=233(33)															
	259.6	141.1	65.92	86.63	82.20	31.89	48.27	65.41	41.46	66.46	40.91	19.56	14.57	25.92	13.82
	235.0	126.0	58.8	76.5	72.2	26.3	40.2	55.1	31.0	57.0	33.0	14.5	10.9	22.2	11.4
	293.0	164.0	71.8	98.4	94.0	39.1	55.0	76.9	49.0	78.5	50.8	24.1	17.2	30.5	16.7
	12.27	6.11	2.46	4.17	3.76	2.09	2.73	4.72	3.10	4.56	2.96	1.53	1.21	1.42	.99
	4.73	4.33	3.73	4.81	4.57	6.55	5.66	7.22	7.45	6.86	7.24	7.82	8.30	5.48	7.16
Total sample of northern and western *C. lupus*, female, n=146(33)															
	247.7	133.5	64.89	83.70	78.30	30.53	45.73	61.35	40.46	62.68	38.50	18.23	13.53	24.79	13.44
	224.0	120.0	59.2	73.7	70.2	24.3	39.3	50.6	34.1	54.0	32.4	14.8	11.4	22.2	11.2
	278.0	154.0	71.3	95.1	90.3	37.6	53.9	73.5	48.5	75.8	45.8	23.4	15.9	28.2	16.3
	12.18	6.62	2.69	3.78	3.60	2.10	2.59	4.18	2.86	4.26	2.82	1.35	.97	1.26	.91
	4.92	4.96	4.14	4.52	4.56	6.88	5.66	6.81	7.07	6.80	7.32	7.40	7.17	5.08	6.77
Total sample of northern and western *C. latrans*, male, n=166(2)															
	197.1	99.4	57.79	70.04	56.53	20.06	30.91	46.77	34.10	46.96	26.06	12.15	9.35	20.38	11.81
	178.0	88.0	52.0	61.4	49.7	17.5	27.1	36.1	29.1	39.5	21.2	9.1	7.7	17.6	9.7
	213.0	109.0	63.6	78.4	62.4	29.7	35.5	54.6	39.7	52.8	29.6	14.3	11.0	22.8	13.8
	7.28	3.96	2.00	2.85	2.40	1.14	1.40	3.13	2.13	2.61	1.63	.88	.62	.96	.64
	3.69	3.98	3.46	4.07	4.24	5.68	4.53	6.69	6.25	5.56	6.25	7.24	6.63	4.71	5.42

	1	2	3	4	5	6	7	8	9	10	11	12	13	14	15

Total sample of northern and western *C. latrans*, female, n=111(4)

188.1	95.0	56.80	67.39	54.19	19.47	29.44	44.81	33.74	44.49	24.75	11.68	8.78	19.60	11.52
172.0	87.0	53.7	60.7	48.8	16.7	26.7	38.7	28.7	39.0	21.9	9.7	7.7	17.7	10.4
204.0	106.0	60.2	74.0	60.4	22.5	32.4	53.6	40.9	51.3	28.7	13.8	10.0	21.6	13.0
6.57	3.45	1.59	2.56	2.22	1.12	1.55	3.12	2.18	2.36	1.46	.86	.56	.86	.55
3.49	3.63	2.80	3.80	4.10	5.75	5.26	6.96	6.46	5.30	5.90	7.36	6.38	4.39	4.77

3. *C. lupus lycaon*, western group, male, n=42(11)

253.2	136.7	66.38	84.39	79.42	30.31	45.75	64.52	41.56	65.70	38.65	18.72	13.61	25.19	14.02
238.0	125.0	59.8	77.5	73.0	26.7	41.6	54.1	35.3	60.0	34.0	16.3	11.8	23.7	12.8
274.0	150.0	72.0	90.3	85.5	34.3	50.6	76.5	46.3	76.0	45.0	22.8	15.5	27.4	15.7
7.35	5.00	2.99	2.71	2.80	1.93	2.10	5.61	2.57	3.25	2.35	1.27	.99	.93	.67
2.90	3.66	4.51	3.21	3.52	6.38	4.59	8.69	6.18	4.95	6.07	6.80	7.30	3.70	4.79

C. lupus lycaon, western group, female, n=30(8)

241.2	129.0	64.34	81.07	75.14	28.81	42.47	60.43	40.14	61.21	36.04	17.24	12.46	23.56	13.39
224.0	121.0	59.0	72.4	70.5	26.0	39.5	49.4	34.9	51.7	32.0	15.6	10.4	21.1	11.0
268.0	142.0	69.3	87.5	81.0	32.5	47.3	89.7	46.1	70.0	40.9	20.8	13.9	26.0	14.8
10.03	5.36	2.67	3.34	2.86	1.64	1.79	4.81	2.76	4.26	2.20	1.20	.89	1.27	.83
4.16	4.15	4.16	4.12	3.81	5.71	4.22	7.96	6.88	6.96	6.10	6.98	7.18	5.38	6.18

C. lupus lycaon, eastern group, male, n=19(4)

247.1	134.1	63.25	82.69	77.85	28.59	44.26	60.99	39.78	63.34	37.78	18.15	13.09	24.55	14.25
237.0	128.0	58.3	78.5	74.2	26.0	40.8	49.4	36.0	59.0	34.3	16.3	12.0	22.6	13.4
255.0	140.0	68.0	87.7	84.3	32.0	47.6	72.8	44.9	69.5	42.2	20.0	14.8	27.5	15.7
5.96	3.59	2.49	2.48	2.71	1.47	1.69	5.52	2.77	2.86	1.89	1.08	.80	1.20	.66
2.41	2.68	3.93	2.99	3.48	5.14	3.81	9.07	6.95	4.51	5.00	5.93	6.12	4.88	4.64

C. lupus lycaon, eastern group, female, n=12(2)

231.4	125.0	62.68	79.15	73.76	26.39	40.76	56.58	36.93	59.32	35.10	16.54	11.98	22.67	13.64
223.0	116.0	60.5	75.0	69.0	23.5	37.6	51.9	35.0	54.6	32.3	14.0	11.3	21.3	13.2
241.0	132.0	66.0	83.5	78.3	30.0	44.3	60.6	42.5	64.4	37.7	18.5	12.7	24.2	14.3
6.64	4.79	1.70	2.92	3.20	2.17	2.08	2.86	2.21	3.30	1.96	1.49	.50	.93	.35
2.87	3.82	2.71	3.69	4.34	8.21	5.11	5.05	5.99	5.57	5.58	9.01	4.21	4.10	2.61

4. Early specimens of *C. rufus*, as listed in table 2, male, n=11(4)*

235.0	121.0	64.1	84.0	70.0	25.1	38.3	56.7	41.3	59.9	34.5	16.3	11.1	24.5	13.1
250.0	128.0	65.4	81.0	71.1	26.6	43.8	61.9	43.3	63.8	37.6	18.3	12.6	25.2	14.2
235.0	126.0	60.6	78.5	71.0	27.2	42.3	57.4	37.7	57.5	35.3	18.4	13.7	23.8	13.5
228.0	115.0	64.1	79.5	66.0	24.0	39.0	58.9	41.3	57.5	34.3	18.3	11.2	23.5	14.4
246.0	120.0	61.1	82.5	70.0	28.5	42.5	53.0	35.4	60.3	35.2	15.8	13.6	24.8	13.9
240.0	122.0	64.1	78.9	72.5	26.7	40.4	52.4	35.7	62.7	34.2	15.8	12.2	23.6	13.7
234.0	114.0	61.1	80.4	70.7	27.4	39.9	49.8	38.5	61.7	34.0	16.8	12.4	23.8	14.4
224.0	122.0	60.9	77.5	70.1	28.1	40.5	57.6	38.3	56.3	33.2	16.0	11.4	24.6	13.8
250.0	126.0	62.6	82.8	71.2	25.8	39.4	58.2	32.3	60.6	32.2	15.4	11.1	23.8	14.6
247.0	124.0	65.7	82.6	71.0	26.1	40.2	56.5	37.0	62.7	35.0	16.2	11.2	24.8	15.6
240.0	120.0	66.4	80.2	69.4	26.7	40.2	52.5	38.3	62.0	32.5	17.5	13.5	24.5	13.1

Early specimens of *C. rufus*, as listed in table 2, female, n=3(1)*

222.0	114.0	60.5	75.0	66.0	24.4	37.3	48.4	31.5	55.7	33.6	15.2	10.5	22.1	12.0
230.0	117.0	60.7	79.5	65.0	25.5	38.8	54.0	38.3	58.4	32.3	13.7	10.1	22.3	13.0
222.0	115.0	64.1	77.3	64.6	25.5	37.0	52.3	37.3	54.5	31.3	14.5	11.3	22.2	13.4

5. *C. rufus gregoryi*, south-central United States, 1919-1929, male, n=63(1)

232.6	121.2	61.93	78.98	69.37	26.21	40.00	55.52	37.33	58.32	33.64	15.78	11.93	23.66	13.68
218.0	110.0	58.3	72.6	63.6	22.3	35.7	47.2	32.0	51.8	29.1	13.3	10.5	21.4	10.6
261.0	138.0	68.0	86.8	75.3	32.0	47.2	62.1	42.5	65.4	38.0	18.5	13.2	26.0	16.0
8.76	5.93	2.10	2.79	2.73	1.95	2.24	3.73	2.40	3.45	2.24	1.26	.73	1.00	.82
3.77	4.89	3.39	3.53	3.93	7.45	5.61	6.71	6.43	5.91	6.65	7.96	6.10	4.23	5.99

	1	2	3	4	5	6	7	8	9	10	11	12	13	14	15
C. rufus gregoryi, south-central United States, 1919-1929, female, n=52(1)															
	220.9	115.4	61.14	75.15	66.78	25.32	37.99	52.93	37.93	53.89	31.66	14.84	11.12	22.31	13.29
	210.0	108.0	57.6	68.5	61.6	21.2	33.4	42.7	30.8	50.4	27.3	12.0	9.6	20.0	11.7
	245.0	130.0	64.8	80.5	74.7	29.7	45.0	63.0	41.5	66.1	36.1	17.3	12.9	24.4	14.7
	5.58	4.40	1.89	2.58	2.85	1.98	2.23	4.20	2.44	2.66	1.72	1.02	.71	1.08	.72
	2.53	3.81	3.11	3.45	4.27	7.62	5.87	7.94	6.42	4.93	5.33	6.88	6.36	4.84	5.42
6. Specimens from inland eastern Texas, post-1960, male, n=77(6)															
	206.8	104.2	58.61	71.90	59.22	21.56	33.32	49.55	35.57	50.74	27.91	13.15	10.05	21.04	12.29
	192.0	96.0	54.6	65.5	53.3	18.8	29.5	43.5	29.6	44.0	24.3	10.8	8.8	19.2	10.3
	221.0	112.0	63.7	76.8	64.8	24.6	36.3	61.5	40.2	56.8	30.5	15.1	11.5	23.0	13.6
	5.62	3.45	2.11	2.14	1.91	1.25	1.42	3.59	2.12	2.64	1.44	.97	.56	.85	.64
	2.72	3.31	3.60	2.98	3.22	5.80	4.26	7.24	5.96	5.21	5.17	7.35	5.56	4.02	5.23
Specimens from inland eastern Texas, post-1960, female, n=42(6)															
	198.0	99.52	58.01	69.81	57.39	20.59	31.75	47.36	35.66	47.86	26.60	12.41	9.51	20.43	12.31
	180.0	91.0	54.2	64.7	53.6	18.6	29.1	42.1	30.1	42.2	22.3	10.7	8.4	18.5	10.5
	214.0	109.0	62.0	75.5	62.8	23.1	37.8	56.0	39.8	54.8	30.8	15.5	11.5	23.0	13.6
	7.51	3.78	1.61	2.77	1.94	1.08	1.65	3.23	2.01	3.20	1.77	1.01	.64	.96	.72
	3.80	3.79	2.77	3.97	3.39	5.24	5.19	6.83	5.63	6.69	6.65	8.15	6.70	4.72	5.85
7. Specimens from Jefferson and eastern Chambers counties, Texas, 1963-1970, male, n=15(3)															
	233.7	119.1	60.63	76.73	68.05	26.59	39.59	50.99	35.79	60.23	33.31	15.07	11.31	22.27	13.86
	218.0	105.0	56.0	67.0	61.5	24.0	33.8	41.3	27.5	52.5	28.3	11.8	10.0	21.1	12.0
	247.0	130.0	64.7	82.6	73.6	29.2	43.0	58.0	40.0	65.0	36.8	16.7	12.5	23.3	14.8
	6.98	5.95	2.39	3.76	3.06	1.71	2.31	4.14	3.44	3.23	2.30	1.23	.64	.79	.80
	2.99	4.99	3.95	4.90	4.50	6.43	5.83	8.12	9.60	5.36	6.90	8.15	5.63	3.55	5.79
Specimens from Jefferson and eastern Chambers counties, Texas, 1963-1970, female, n=4*															
	222.0	112.0	58.7	75.0	65.5	27.7	38.5	47.7	33.5	55.0	33.6	15.5	11.0	21.1	12.3
	224.0	111.0	61.8	77.1	63.0	24.0	35.8	52.5	39.1	56.7	33.7	14.7	10.6	21.1	13.2
	220.0	111.0	58.8	76.1	66.8	25.4	35.4	48.5	35.7	58.0	30.8	14.0	10.1	21.1	13.9
	225.0	110.0	59.6	74.0	63.8	24.1	38.3	45.8	33.8	55.5	31.6	15.3	10.8	21.2	12.5
8. *C. cedazoensis*, Cedazo, Aguascalientes															
	----	----	----	----	----	----	----	----	----	----	----	----	----	16.9	----
9. *C. lepophagus*, Hagerman, Idaho															
	----	----	----	----	----	----	----	----	----	----	----	----	----	17.0	10.3
	----	----	----	----	----	----	----	----	----	----	----	----	----	16.0	10.0
C. lepophagus, Rexroad fauna, Kansas															
	----	----	----	----	----	----	----	----	----	----	----	----	----	19.1	----
	----	----	----	----	----	----	----	----	----	----	----	----	----	16.9	----
	----	----	----	----	----	----	----	----	----	----	----	----	----	----	10.2
C. lepophagus, Broadwater, Nebraska															
	----	----	----	----	----	----	----	----	----	----	----	----	----	19.1	11.5
	----	----	----	----	----	----	----	----	----	----	----	----	----	19.6	12.3
C. lepophagus, Lisco, Nebraska															
	----	----	----	----	----	----	----	----	----	----	----	----	----	17.9	11.0
C. lepophagus, Cita Canyon, Texas															
	194.0	103.0	----	68.6	----	20.5	----	54.0	----	----	----	11.8	----	----	----
	187.0	----	54.3	64.7	----	----	----	----	----	----	26.6	----	8.7	19.0	11.3
	----	----	----	----	----	----	----	57.2	35.0	51.0	----	----	----	----	----
	190.0	99.0	50.0	70.0	----	20.0	----	48.0	33.0	----	25.0	12.0	----	20.0	12.0
	----	----	----	74.5	----	----	----	----	----	----	----	----	8.8	20.7	12.0

	1	2	3	4	5	6	7	8	9	10	11	12	13	14	15
10.	*C. latrans*, Papago Springs Cave, Arizona															
	174.0	99.0	56.9	67.1	57.1	20.2	30.0	40.6	32.9	43.1	25.4	12.6	----	19.6	11.3
	C. latrans, McKittrick, California															
	----	98.6	58.3	74.5	59.5	----	----	47.0	38.2	----	25.2	11.5	----	22.0	12.6
	----	----	62.4	76.5	65.0	24.5	----	----	----	----	27.6	----	----	22.3	12.3
	210.0	----	----	77.0	64.4	22.5	35.5	55.0	36.0	----	28.0	13.1	----	22.5	12.6
	219.0	110.0	58.4	77.5	64.5	21.5	----	53.1	35.5	----	28.8	14.0	----	24.0	13.8
	207.0	----	----	72.5	61.1	23.0	33.3	----	----	----	----	----	10.3	21.3	12.3
	206.0	----	61.5	75.5	61.8	20.0	33.5	56.0	36.5	----	----	----	----	22.5	----
	C. latrans, Maricopa, California															
	204.0	----	59.0	73.5	56.2	21.8	32.8	----	----	----	----	----	----	20.1	----
	C. latrans, Rancho La Brea, California															
	n=44	n=36	n=42	n=49	n=21	n=44	n=38	n=44	n=47	n=40	n=40	n=36	n=8	n=43	n=33
	205.5	106.7	60.54	72.54	61.20	22.18	34.29	51.21	36.73	48.16	27.05	13.07	9.88	21.05	11.84
	185.0	90.0	55.7	65.5	50.0	17.3	29.8	44.0	32.9	44.0	22.0	10.2	9.0	18.2	10.2
	222.0	116.0	65.1	78.0	67.4	25.5	38.3	61.5	42.0	52.8	32.3	15.2	10.9	23.5	13.6
	9.03	5.58	1.76	3.23	4.31	1.70	2.00	4.14	2.12	2.55	2.33	1.10	.53	1.29	.89
	4.39	5.23	2.91	4.45	7.05	7.68	5.83	8.08	5.78	5.30	8.63	8.41	5.32	6.14	7.54
	C. latrans, Rancho La Brea, California (unusually small specimen)															
	179.0	90.0	56.5	64.0	50.8	19.6	28.7	41.1	34.9	41.8	22.5	10.2	----	18.0	10.2
	Type of *C. andersoni*															
	173.0	93.0	57.5	63.5	56.0	----	35.1	38.3	33.3	40.0	24.5	12.2	----	20.5	----
11.	*C. edwardii*, Curtis Ranch, Arizona															
	----	118.0	----	77.0	65.4	22.2	----	----	----	----	29.1	13.7	12.4	24.0	12.6
	C. edwardii, Rome Beds, Oregon															
	----	----	59.0	78.8	68.0	19.5	----	50.8	38.8	----	----	----	11.3	24.0	13.7
12.	*C. rufus*, Eddy Bluff shelter, Arkansas															
	----	----	----	----	----	----	----	----	----	----	----	----	----	23.0	13.1
	C. rufus, Haile VIIA, Florida															
	----	----	----	78.0	----	----	----	----	----	----	----	14.5	----	20.9	13.3
	C. rufus, Inglis IA, Florida															
	----	----	----	----	----	----	----	----	----	----	----	----	----	22.1	----
	----	----	----	----	----	----	----	----	----	----	----	----	----	22.8	----
	C. rufus, Port Kennedy, Pennsylvania															
	----	----	----	----	----	----	----	----	----	----	----	----	----	24.0	12.3
13.	*C. armbrusteri*, McCleod, Florida															
	----	----	----	----	----	----	----	----	----	----	----	----	----	26.4	14.9
	C. armbrusteri, Coleman IIA, Florida															
	----	133.0	66.4	85.6	78.2	24.0	----	63.0	45.2	----	35.0	18.5	----	26.0	15.5
	----	----	----	----	----	----	----	----	----	----	----	----	----	26.1	13.9
	----	----	----	----	----	----	----	----	----	----	----	----	----	28.8	----
	----	----	----	----	----	----	----	----	----	----	----	----	----	26.5	----

	1	2	3	4	5	6	7	8	9	10	11	12	13	14	15

C. armbrusteri, Cumberland Cave, Maryland

1	2	3	4	5	6	7	8	9	10	11	12	13	14	15
258.0	128.0	----	96.0	82.2	27.0	43.5	61.5	50.0	61.7	34.5	17.2	----	28.0	----
----	----	64.8	----	78.1	----	----	59.8	44.5	51.0	----	----	----	26.6	14.1
----	----	----	----	----	----	----	----	----	----	----	----	10.5	----	----
----	150.0	----	95.0	87.0	29.0	45.8	68.0	44.0	----	----	----	12.8	28.7	15.7
285.0	161.0	75.0	98.0	80.0	30.7	49.0	70.0	43.0	77.0	45.0	21.0	----	27.9	15.5
----	----	----	----	----	----	----	----	----	----	----	----	----	28.9	15.2
270.0	----	----	94.0	----	----	----	----	----	65.0	35.0	----	----	28.0	17.0
----	----	65.0	----	83.5	----	----	59.0	39.8	60.0	35.6	19.0	----	29.5	17.1
----	----	----	----	----	----	----	----	----	----	----	----	----	28.9	15.2

C. armbrusteri, Rushville, Nebraska

1	2	3	4	5	6	7	8	9	10	11	12	13	14	15
----	----	----	----	----	----	----	----	----	----	----	----	----	30.0	15.5

14. *C. lupus*, Hunker Creek, Yukon

1	2	3	4	5	6	7	8	9	10	11	12	13	14	15
257.0	----	61.0	88.7	----	35.8	52.0	57.5	40.0	66.3	42.7	20.0	----	28.5	13.7

C. lupus, Maricopa, California

1	2	3	4	5	6	7	8	9	10	11	12	13	14	15
258.0	144.0	69.0	87.5	----	33.8	48.2	64.0	39.0	----	41.6	18.8	----	26.6	13.7
255.0	----	----	----	----	32.5	----	----	----	----	----	----	----	26.6	12.5
----	----	----	----	----	----	----	----	----	----	----	----	----	26.2	----
----	----	----	----	----	----	----	----	----	----	----	----	----	28.5	----
----	----	----	----	----	----	----	----	----	----	----	----	----	23.3	13.1
----	----	----	----	----	----	----	----	----	----	----	----	----	28.1	12.2
----	135.0	----	----	----	----	----	----	----	----	----	----	----	----	----

C. lupus, Rancho La Brea, California

1	2	3	4	5	6	7	8	9	10	11	12	13	14	15
240.0	138.0	62.7	81.5	79.7	33.6	48.1	74.3	46.0	62.7	38.0	25.0	----	26.8	14.1
----	140.0	----	82.7	80.3	28.5	44.0	64.0	44.0	----	39.0	18.3	----	24.8	14.2
----	123.0	66.9	78.7	72.3	29.0	----	58.9	42.2	56.0	35.6	16.0	----	23.0	12.5
230.0	128.0	66.0	77.8	78.8	29.0	45.0	53.0	42.7	56.3	37.0	19.0	----	26.8	13.1
265.0	150.0	71.4	90.0	85.7	33.3	49.4	68.0	48.2	66.7	39.5	21.0	----	29.0	14.3
269.0	147.0	69.2	89.9	----	36.1	52.0	75.0	41.3	70.5	44.5	22.1	----	26.8	13.2
252.0	----	62.7	82.5	79.9	33.5	44.6	----	----	----	63.8	39.5	19.7	25.1	13.8
----	148.0	69.0	94.0	94.5	38.0	----	68.3	43.7	70.6	44.3	23.5	----	29.2	12.8
248.0	140.0	64.5	84.0	83.2	31.0	46.8	57.5	41.5	62.0	----	----	----	27.6	13.9
----	----	----	----	----	----	----	----	----	----	----	----	----	23.5	12.5
----	----	----	----	----	----	----	----	----	----	----	----	----	25.5	----

Type of *C. milleri*

1	2	3	4	5	6	7	8	9	10	11	12	13	14	15
247.0	137.0	66.7	83.0	89.6	33.5	49.8	63.9	44.3	58.6	39.0	18.8	----	28.6	13.2

C. lupus, Goodland, Kansas

1	2	3	4	5	6	7	8	9	10	11	12	13	14	15
223.1	126.0	67.0	76.5	73.0	30.0	43.6	59.0	34.0	54.5	34.0	15.0	13.3	23.1	14.0

C. lupus, Hay Springs, Nebraska

1	2	3	4	5	6	7	8	9	10	11	12	13	14	15
----	----	----	----	----	----	----	----	----	----	----	----	----	27.4	13.6

C. lupus, Hermit's Cave, New Mexico

1	2	3	4	5	6	7	8	9	10	11	12	13	14	15
----	----	----	----	----	----	----	----	----	----	----	----	14.9	27.9	14.6

C. lupus, San Josecito Cave, Nuevo Leon

1	2	3	4	5	6	7	8	9	10	11	12	13	14	15
216.0	125.0	63.0	76.5	75.0	26.5	40.0	54.0	40.5	50.5	32.4	17.7	11.9	25.0	12.4

15. *C. dirus*, McKittrick, California

1	2	3	4	5	6	7	8	9	10	11	12	13	14	15
309.0	170.0	72.5	106.5	99.0	40.0	58.0	78.5	47.8	75.5	49.4	24.0	15.8	33.0	15.6

	1	2	3	4	5	6	7	8	9	10	11	12	13	14	15
C. dirus, Maricopa, California															
	311.0	174.0	76.0	106.1	----	44.3	69.0	90.8	52.5	77.5	47.0	27.0	----	----	----
	298.0	----	----	100.0	96.7	38.0	59.0	93.0	----	----	45.2	24.5	17.2	32.3	----
	300.0	----	----	102.5	----	39.0	60.0	----	----	----	43.8	22.5	----	----	----
	----	160.0	74.5	100.0	----	40.0	----	82.5	47.9	75.0	43.0	21.5	----	31.8	14.9
	310.0	168.0	----	101.0	----	40.0	65.0	95.0	54.0	----	----	22.0	----	----	16.0
	318.0	----	80.0	105.0	95.2	39.0	58.0	87.0	----	----	45.0	21.5	----	32.5	15.4
	290.0	158.0	73.0	101.0	97.0	38.0	----	89.0	47.0	----	----	22.8	----	31.0	14.4
C. dirus, Rancho La Brea, California, n=62															
	294.8	163.3	74.73	99.99	96.15	39.27	58.02	83.45	49.33	72.43	42.39	21.75	15.66	31.75	15.15
	258.0	148.0	64.0	85.0	87.7	35.0	52.0	73.4	43.5	63.5	36.6	18.8	13.5	28.7	13.1
	316.0	177.0	83.0	107.0	104.0	45.3	65.5	100.0	54.4	77.5	48.5	26.5	17.5	35.3	17.0
	11.31	7.15	3.08	4.18	3.92	2.36	3.30	5.63	2.13	3.54	2.66	1.66	1.15	1.38	.90
	3.84	4.38	4.12	4.17	4.08	6.01	5.68	6.75	4.32	4.89	6.38	7.63	7.34	4.35	5.94
C. dirus, Hornsby Springs, Florida															
	----	----	----	----	----	----	----	----	----	----	----	----	----	29.6	14.5
	----	----	----	----	----	----	----	----	----	----	----	----	----	32.2	----
C. dirus, Reddick IA, Florida															
	----	----	----	----	----	----	----	----	----	----	----	----	----	30.7	16.5
	----	----	----	----	----	----	----	----	----	----	----	----	----	30.8	15.5
C. dirus, Melbourne, Florida															
	----	----	----	----	----	----	----	----	----	----	----	----	----	32.0	----
	----	----	----	----	----	----	----	----	----	----	----	----	----	30.0	----
C. dirus, Bradenton, Florida															
	----	----	----	----	----	----	----	----	----	----	----	----	----	30.1	15.1
C. dirus, Ohio River, Indiana															
	----	----	----	----	----	----	----	----	----	----	----	----	----	35.5	14.0
C. dirus, Twelve Mile Creek, Kansas															
	----	----	----	----	----	----	----	----	----	----	----	----	----	30.0	15.4
C. dirus, Welsh Cave, Kentucky															
	309.0	180.0	78.0	104.3	100.7	39.0	59.1	104.0	57.3	82.0	45.5	25.0	----	30.5	15.7
C. dirus, Herculaneum, Missouri															
	----	----	----	----	----	----	----	----	----	----	----	----	----	32.5	15.6
C. dirus, Hermit's Cave, New Mexico															
	----	----	----	----	----	----	----	----	----	----	40.4	20.7	----	33.0	15.4
C. dirus, Marlow, Oklahoma															
	310.0	170.0	78.5	111.5	102.5	37.0	60.0	100.0	50.0	79.0	49.0	24.0	18.0	31.0	----
C. dirus, Ingleside, Texas															
	333.0	179.0	79.0	110.0	----	44.0	----	----	53.0	----	48.8	27.0	----	35.5	17.5
C. dirus, San Josecito Cave, Nuevo Leon															
	----	161.0	72.0	102.0	101.2	37.6	----	----	49.2	----	45.5	21.4	----	33.6	15.4
	297.0	169.0	76.0	104.3	103.0	37.2	58.8	94.0	54.2	74.0	44.9	21.0	14.8	33.7	15.0

APPENDIX C

The numbers along the top of each of the following columns correspond to four measurements of the mandible and lower dentition: (1) distance from anterior edge of alveolus of p1 to posterior edge of alveolus of m3; (2) minimum depth from dorsal surface of mandible between p3 and p4 to ventral surface of mandible; (3) crown length of p4; (4) crown length of m1. If a sample size (n) is listed for a series, then the five horizontal rows under that figure are mean, lower extreme, upper extreme, standard deviation, and coefficient of variation. If no sample size is given, the numbers shown are actual measurements of individuals, rather than means, etc.

	1	2	3	4
1. *C. lepophagus*				
Santa Fe River, Florida	78.0	15.9	12.9	22.5
	83.5	19.4	12.3	22.6
	----	20.0	----	----
Grand View, Idaho	82.5	19.1	----	20.8
	----	18.7	13.5	22.4
Hagerman, Idaho	----	----	----	19.0
	----	17.1	13.2	----
	70.5	17.5	12.3	20.8
Rexroad fauna, Kansas	70.0	14.0	11.5	18.5
	78.7	16.0	----	----
	----	----	----	18.1
Broadwater, Nebraska	75.5	----	----	20.5
Lisco, Nebraska	70.0	14.5	11.1	17.9
Cita Canyon, Texas	n=14	n=14	n=16	n=13
	78.61	18.16	13.08	21.21
	73.2	15.1	12.1	19.5
	82.0	21.3	14.0	23.0
	2.63	1.48	.52	1.15
	3.35	8.14	3.94	5.42
2. *C. latrans*				
Recent, western U.S., male	n=99	n=99	n=99	n=99
	79.78	17.10	12.58	21.94
	71.4	14.3	10.7	19.5
	88.5	19.9	14.0	24.3
	3.26	1.14	.70	1.01
	4.10	6.66	5.58	4.60
Recent, western U.S., female	n=99	n=99	n=99	n=99
	76.35	15.98	12.08	21.10
	69.1	12.7	10.8	18.6
	82.3	19.0	13.9	23.4
	2.71	1.14	.60	.90
	3.55	7.12	4.96	4.26
Irvington, California	79.9	19.3	13.8	23.9
	78.9	16.9	13.4	21.7
McKittrick, California	n=10	n=10	n=14	n=16
	82.24	18.94	12.78	22.78
	77.0	17.4	11.7	21.0
	88.5	20.4	14.1	24.6
	3.82	1.14	.70	1.11
	4.64	6.02	5.51	4.88

	1	2	3	4
Maricopa, California	n=10	n=10	n=15	n=22
	82.80	18.15	12.75	23.13
	80.0	17.0	11.6	21.1
	85.0	19.8	13.8	25.0
	1.81	1.00	.55	.84
	2.19	5.55	4.33	3.62
Rancho La Brea, California	n=41	n=41	n=37	n=40
	81.77	18.69	12.56	22.44
	77.0	16.0	11.3	21.0
	87.5	21.1	13.7	25.2
	2.52	1.30	.62	1.00
	3.08	6.97	4.90	4.45
Vallecito Creek, California	85.5	18.9	13.1	22.0
Haile XIIB, Florida	78.1	16.8	----	----
Devil's Den, Florida	72.1	15.2	----	20.2
	----	17.5	----	21.5
Lake Cutaline, Florida	----	18.0	11.7	----
Rushville, Nebraska	----	----	----	21.4
Mullen, Nebraska	73.6	17.5	11.6	20.4
Frankstown Cave, Pennsylvania	83.5	19.6	13.7	23.3
Lewisville, Texas	82.0	17.8	13.8	23.8
Friesenhahn Cave, Texas	78.6	18.0	----	21.4
	75.5	16.0	12.3	20.4
	----	15.8	12.1	----
	76.5	15.5	11.4	20.7
San Josecito Cave, Nuevo Leon	n=12	n=15	n=10	n=9
	80.25	18.01	12.96	22.44
	76.0	16.1	12.0	20.8
	84.0	19.9	14.1	24.8
	2.61	1.24	.69	1.44
	3.25	6.90	5.30	6.43

3. *C. edwardii*

	1	2	3	4
Anita, Arizona	96.5	19.5	16.0	27.5
Curtis Ranch, Arizona	87.2	20.0	15.2	24.8
Miñaca Mesa, Chihuahua	91.5	----	14.5	23.3
Hemphillian specimen from Ash Hollow formation, Nebraska	82.0	20.3	13.5	25.0

4. *C. rufus*

	1	2	3	4
Recent, male	n=64	n=64	n=64	n=64
	89.66	21.60	14.40	25.68
	83.2	19.2	13.2	23.4
	99.5	24.9	16.0	28.1
	3.37	1.31	.80	1.04
	3.77	6.08	5.56	4.05
Recent, female	n=61	n=61	n=61	n=61
	86.05	20.98	13.94	24.52
	79.6	18.1	12.3	22.1
	93.6	25.2	15.4	27.1
	2.78	1.47	.72	1.11
	3.24	7.00	5.16	4.54

	1	2	3	4
Inglis IA, Florida	----	----	----	27.5
	87.5	20.7	15.6	26.0
Melbourne, Florida	94.0	25.2	15.0	26.3
Port Kennedy, Pennsylvania	----	----	15.8	----
5. *C. armbrusteri*				
Anita, Arizona	105.5	27.5	18.0	----
McCleod, Florida	102.5	23.6	16.4	31.3
	113.0	27.5	17.8	32.0
Coleman IIA, Florida	----	----	----	27.4
	----	----	----	28.5
	----	----	----	30.9
	----	----	----	28.0
	----	----	----	32.0
Cumberland Cave, Maryland	104.0	27.0	17.3	29.4
	----	----	18.2	----
	110.0	28.2	17.6	30.5
	106.5	28.2	18.4	30.0
	105.0	28.7	17.7	30.5
	----	----	----	30.7
	----	----	----	30.9
	----	----	----	30.9
6. *C. lupus*				
Recent, western U.S., male	$n=62$	$n=62$	$n=62$	$n=62$
	95.31	26.62	15.68	28.53
	86.7	23.0	13.6	26.0
	104.0	31.0	17.0	31.5
	3.44	1.66	.73	1.20
	3.61	6.25	4.68	4.22
Recent, western U.S., female	$n=47$	$n=47$	$n=47$	$n=47$
	92.15	24.90	15.04	27.07
	84.7	22.5	12.7	25.1
	97.5	27.7	16.5	30.0
	2.66	1.29	.74	1.00
	2.88	5.17	4.95	3.70
Maricopa, California	----	----	17.5	30.2
	----	27.5	16.3	29.1
	----	----	15.5	28.4
	97.7	27.5	16.1	30.5
Rancho La Brea, California	99.9	29.1	18.2	31.2
	95.5	26.8	16.8	29.0
Type of *C. milleri*	95.0	25.8	18.0	32.0
Goodland, Kansas	84.3	22.1	15.0	26.3
Millington, Michigan	101.0	28.0	15.0	30.0
Hay Springs, Nebraska	98.0	29.0	17.4	29.9
Mullen, Nebraska	100.0	30.0	17.3	29.3
Hermit's Cave, New Mexico	96.0	30.6	16.6	30.2
	95.0	24.6	15.8	29.9
San Josecito Cave, Nuevo Leon	84.0	23.0	14.3	26.5

	1	2	3	4
7. *C. dirus*				
Murray Springs, Arizona	110.0	----	18.6	34.8
	----	----	18.4	33.9
Cool quarry, California	----	31.5	19.0	----
Teichart gravel pit, California	----	----	20.8	----
Arroyo Las Positas, California	110.0	31.9	19.0	34.3
McKittrick, California	n=7	n=7	n=13	n=15
	113.40	33.73	20.08	35.33
	106.5	32.0	18.8	33.7
	123.0	35.0	21.7	37.0
	5.57	1.25	.89	.98
	4.91	3.71	4.41	2.78
Maricopa, California	n=10	n=12	n=18	n=17
	113.2	34.01	19.87	35.01
	108.0	29.8	18.0	33.5
	119.5	36.9	21.5	37.5
	3.82	2.38	1.04	1.47
	3.36	7.01	5.23	4.19
Rancho La Brea, California	n=73	n=73	n=73	n=73
	110.64	31.81	19.48	34.25
	102.0	25.5	17.9	31.8
	117.5	36.5	20.6	38.5
	3.31	1.96	.66	1.44
	2.99	6.17	3.36	4.20
Ichetucknee River, Florida	117.0	32.5	20.3	35.5
	117.0	31.1	19.8	36.0
	116.5	33.2	----	----
Hornsby Springs, Florida	116.0	29.0	18.3	35.5
Reddick IA, Florida	----	----	19.3	34.4
	----	----	21.3	37.5
Eichelberger Cave, Florida	125.0	32.1	20.8	37.7
	----	34.2	20.9	37.6
Melbourne, Florida	112.5	31.6	19.3	35.3
	----	----	----	33.5
	----	----	----	34.8
	----	----	----	35.9
Bradenton, Florida	----	32.5	18.4	36.2
Twelve Mile Creek, Kansas	----	----	18.0	----
Pendennis, Kansas	----	----	----	34.0
Cragin Quarry, Kansas	108.5	30.5	18.3	----
Welsh Cave, Kentucky	113.5	29.6	19.6	36.3
Conkling Cavern, New Mexico	----	32.1	20.4	35.2
Hermit's Cave, New Mexico	118.0	----	----	35.5
Marlow, Oklahoma	120.0	39.0	20.2	36.0
Frankstown Cave, Pennsylvania	108.6	30.0	17.9	32.0
	110.5	29.2	17.7	32.6

	1	2	3	4
Ingleside, Texas	118.0	34.0	20.1	36.0
Rennick, West Virginia	107.0	28.2	18.0	31.8
Cedazo, Aguascalientes	112.0			
		33.0		34.5
		33.5		34.5
San Josecito Cave, Nuevo Leon	n=9	n=10	n=14	n=18
	110.0	33.23	19.52	34.64
	107.0	30.8	17.8	32.7
	118.2	36.0	20.3	36.5
	3.32	1.63	.71	1.00
	3.02	4.92	3.64	2.89